Design of Active Filters, With Experiments

by

Howard M. Berlin

Originally Published as
The Design of Active Filters, With Experiments
by E&L Instruments, Inc.

Howard W. Sams & Co., Inc.
4300 WEST 62ND ST. INDIANAPOLIS, INDIANA 46268 USA

FIRST EDITION
SECOND PRINTING—1979

International Standard Book Number: 0-672-21539-X
Library of Congress Catalog Card Number: 78-56612

Printed in the United States of America.

Preface

This book was primarily written to provide the individual with a simplified approach to the design and experimentation of active filters. To my knowledge, it is the first book that integrates the design of active filters with a wide range of laboratory experiments. For this reason, this book is useful to the experimenter and hobbyist who wants to learn the basics by self-study, or it can easily serve as an addition to any college course on filter theory or linear integrated circuit design.

It is my strong feeling that active filter design is not difficult. This view is held in spite of a number of texts and handbooks that seem to be heavily dependent on difficult mathematical equations and techniques, giving very few numerical examples, if any. In this book, you are not required to remember or master difficult mathematical expressions, but in most cases, rely on several basic, but necessary, tables and graphs that summarize and simplify the particular design. With them, you will be able to design and build a variety of active filters to meet most requirements. The filters presented in this book represent the best overall choice for each type. For the low-pass and high-pass filters, only the Butterworth response is discussed, since it has the best tradeoff between the passband and stopband responses, in addition to being easily understood. In addition, I believe that the practical hands-on approach given in this book is more beneficial than pure theory for the majority of people, eliminating some of the "black magic" that has surrounded active filter design.

A strong attempt has been made to keep the use of mathematical equations to a bare minimum, giving only the essential relationships. However, the derivation of the major equations and design criteria are presented in the Appendix, which I hope will satisfy some of you.

This is another applications book in the Blacksburg Continuing Education Series. If you have worked with the other books in this series you will find that this book is similar in its approach. As with the other books in the series, the reader will find many useful laboratory-type experiments to illustrate and reinforce the concepts involved.

I would like to thank the editors of the Blacksburg Continuing Education Series: David Larsen, Peter Rony, Jon Titus, and Chris Titus for their encouragement, support, and technical advice; and to E&L Instruments, Inc. for their cooperation and encouragement. Finally, I want to thank my wife Judy for her patience and understanding.

HOWARD M. BERLIN

This book is dedicated to the memory of my mother and uncle

Contents

CHAPTER 6

CHAPTER 7

CHAPTER 8

CHAPTER 9

APPENDIX A

APPENDIX B

APPENDIX C

APPENDIX D

CHAPTER 1

Introduction

This book is about the design of and experiments with active filters. It is the first book to integrate filter design with a series of appropriate problems and experiments. You are not required to perform the experiments; however, as with any subject, I am convinced that one really learns by *doing*, an effective method of reinforcement. Consequently, I have found the practical "hands-on" approach to be the key to understanding active filters.

WHAT IS A FILTER?

> "*A filter is a device or substance that passes electric currents at certain frequencies or frequency ranges while preventing the passage of others.*"—Webster

Specifically, an *active filter* is a device composed of a network of resistors and capacitors built around a solid state device, usually an operational amplifier. On the other hand, *passive filters* are constructed solely by a network of inductors, capacitors, and resistors. This book discusses only active filters, their characteristics and design.

An active filter offers the advantages of:

- No insertion loss.—Since the operational amplifier is capable of providing gain, the input signal will not be immediately attenuated while the filter passes those frequencies of interest.
- Cost.—Active filters on the average will cost far less than passive filters. This is because inductors are expensive and are not always readily available.

- Tuning.—Active filters are easily tuned, or adjusted, over a wide frequency range without altering the desired response.
- Isolation.—As a result of using an operational amplifier, active filters will have a high input and a low output impedance, virtually guaranteeing almost no interaction between the filter and the source or load.

If these characteristics sound too good to be true, I must now be honest with you and tell you that there are some disadvantages or limitations with using active filters. These disadvantages are:

Frequency response.—You are at the mercy of the type of operational amplifier used in your design. This will be outlined later.

Power supply.—Unlike passive filters, active filters require some form of power supply for the operational amplifier.

In the following chapters we shall learn about low-pass, high-pass, bandpass, and notch filters. In addition, you will learn that there is more than one circuit possible for each type.

THE BREADBOARDING SYSTEM

Purpose

The purpose of the breadboarding system used in this text is to give you the practical experience in the wiring and testing of active filters.

Breadboard

The breadboard is designed to accommodate the many experiments that you will perform in this and subsequent chapters. In Fig. 1-1 a top view of the breadboard is shown, which is known as the E&L Instruments, Inc. *SK-10 socket,* and on which you can interconnect the various resistors, capacitors, and operational amplifiers that are required.

Fig. 1-1. Top view of breadboard.

The LR-31 Signal Generator Outboard®*

The LR-31 Signal Generator Outboard® contains an integrated circuit function generator capable of generating sine-waves, square-waves, and triangular waveforms of adjustable frequency with a single control. The LR-31 is powered by an external dual, or bipolar supply, ranging from ±6 to ±12 volts. For simplicity, I have found it easiest to use two heavy-duty 6-volt "lantern" batteries. As shown in Fig. 1-2, the LR-31 is easily mounted on the SK-10 breadboarding socket.

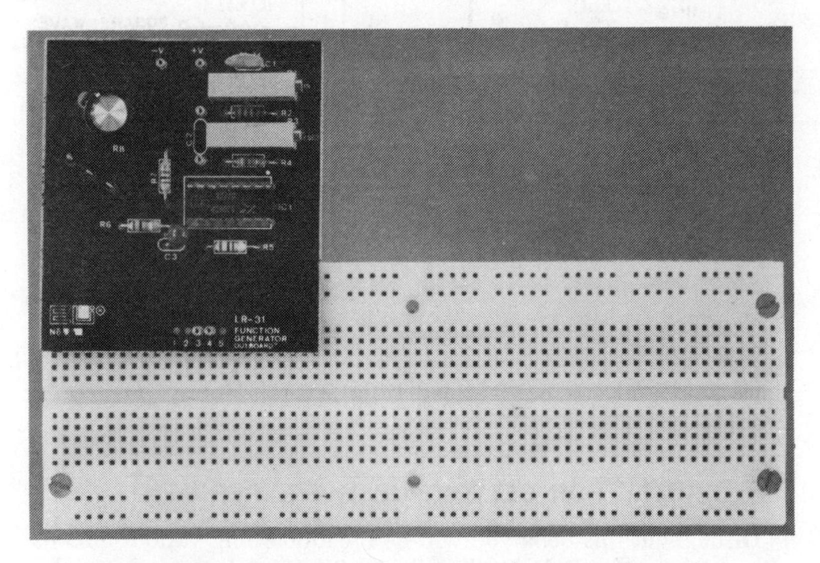

Fig. 1-2. LR-31 Outboard easily mounts on breadboarding socket.

The frequency range of the LR-31 Outboard is determined by inserting an appropriate capacitor into the two socket pins marked "C" on the Outboard, according to the following table:

Frequency Range	*Capacitor*
0.5–1,000 H	1 μF
5–10,000 Hz	.1 μF
50 Hz–100 kHz	.01 μF
500 Hz–1 MHz	.001 μF

As shown in Fig. 1-3, the LR-31 Outboard uses an XR-2206 integrated circuit function generator, made by EXAR Integrated Sys-

* ®E&L Instruments, Inc.

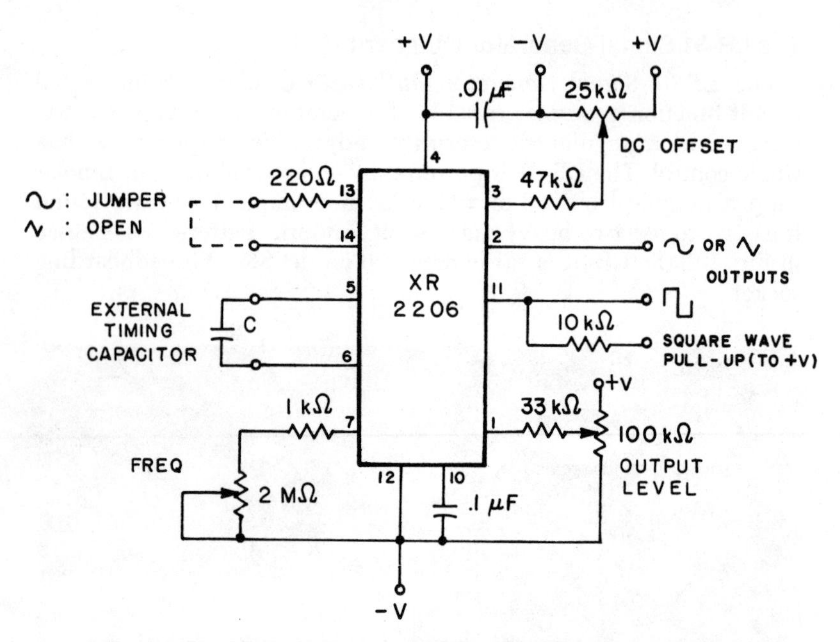

Fig. 1-3. The LR-31 Outboard uses XR-2206 IC function generator.

tems, Inc., and can be obtained from several "hobby" mail order firms.

RULES FOR SETTING UP THE EXPERIMENTS

Throughout this book you will be breadboarding various circuits, either using the LR-31 Outboard and SK-10 socket, or some other method comfortable for you. If you have already had experience with the other Blacksburg Continuing Education Series books, these rules will be familiar. Before you set up any experiment, it is recommended that you do the following:

1. Plan your experiment beforehand. Know what types of results you are expected to observe.
2. Disconnect or turn off *all* power to the breadboard.
3. Clear the breadboard of all wires and components from previous experiments, unless instructed otherwise.
4. Check the wired-up circuit against the schematic diagram to make sure that it is correct.
5. Connect or turn on the power to the breadboard *LAST!*
6. When finished, make sure that you disconnect or turn off all the power to the breadboard *before* you clear the breadboard of wires and components.

FORMAT FOR THE EXPERIMENTS

The instructions for each experiment are presented in the following format.

Purpose

The material presented under this heading states the purpose of your performing the experiment. It is well for you to have this intended purpose in mind as you conduct the experiment.

Schematic Diagram of Circuit

Under this heading is the schematic diagram of the completed circuit that you will wire in the experiment. You should analyze this diagram in an effort to obtain an understanding of the circuit *before* you proceed further.

Design Basics

Under this heading is a summary of the design equations used for the design of the circuit.

Steps

A series of sequential steps describe the instructions for setting up portions of the experiment. Questions are also included at different points of this section. Any numerical calculations are performed easiest on many of the "pocket" type calculators.

HELPFUL HINTS AND SUGGESTIONS

Tools

Only three tools are necessary for all of the experiments given in this book:

1. a pair of "long-nosed" pliers,
2. a wire stripper/cutter,
3. a small screwdriver.

The pliers are used to: straighten out the bent ends of hookup wire that may be used to wire the circuits on the breadboard; straighten out or bend the resistor and capacitor leads to the proper position so that they can be conveniently inserted into the breadboard. The wire stripper/cutter is used to cut the hookup wire to size and to strip about 3/8″ of insulation from each end. The screwdriver is used to adjust the DC OFFSET and OUTPUT LEVEL on the Outboard as instructed in the experiments.

Wire

Only No. 22, No. 24, or No. 26 insulated wire should be used, and it must be solid, not stranded!

Breadboarding

- Never insert too large a wire, or component lead into a breadboard terminal.
- Never insert a bent wire into a breadboard terminal. Straighten out the end with a pair of pliers before inserting it.
- Try to maintain an orderly arrangement of components and wires, keeping all connections as short as possible.

Components and Other Equipment

A variety of fixed resistors and capacitors will be required for the experiments. For those of you who regularly experiment with electronics, your "junk box" should yield most of these. Otherwise, a large number of mail order houses, whose advertisements regularly appear in such magazines as *Popular Electronics*, *Radio-Electronics*, and *Ham Radio*, are possible sources.

	Resistors		*Capacitors*
1 kΩ–1	10 kΩ–7	100 kΩ–1	.001 μF–1
1.5 kΩ–1	12 kΩ–1	180 kΩ–2	.0047 μF–4
2.2 kΩ–1	15 kΩ–1	220 kΩ–2	.01 μF–4
2.7 kΩ–2	18 kΩ–1	270 kΩ–1	.022 μF–2
3 kΩ–1	22 kΩ–1	680 kΩ–1	.033 μF–4
3.3 kΩ–2	27 kΩ–2		.047 μF–2
4.7 kΩ–3	33 kΩ–4		.1 μF–2
5.6 kΩ–2	39 kΩ–1		
6.8 kΩ–7	47 kΩ–2		
8.2 kΩ–4	56 kΩ–2		
	68 kΩ–2		

- *Oscilloscope.* Just about any general purpose oscilloscope will do. A single-trace model is good, but a dual-trace unit is preferred, as it will be useful for rapid comparison of the input and output signals.
- *Function generator.* Any general purpose function generator is fine. These are the ones that have sine-wave, square-wave, and triangle outputs. If you have only an audio signal generator, there will be only a few experiments that you will be unable to perform. As an alternative, you can use the LR-31 Signal Generator Outboard® or construct your own, using the circuit of Fig. 1-3.

- *Pocket calculator.* This item is not mandatory, but it is strongly recommended that you use one. The routine calculations can be accomplished with the simplest of the "4-function" calculators. Believe me, I used to do filter calculations with a slide rule, and the pocket calculator sure saves a lot of time, especially when you can now get one for less than $15.

Operational Amplifier Chips

As there are many types of operational amplifiers available, only one or two will be used for the experiments in this book. These are of the "low-cost" variety that are popular with the majority of hobbyists and scientists with limited funds.

For your experiments, *use only the operational amplifiers that are dual-in-line (DIP) chips, as these are the only ones that are conveniently used with the SK-10 socket.* These chips are similar in appearance to the 14- and 16-pin TTL and CMOS integrated circuit chips.

- *The type-741 Operational Amplifier.* The type-741 operational amplifier is a high performance, general purpose device, which is made by many firms. It is perhaps the most widely used single operational amplifier, costing approximately 35 cents each from several mail order houses. Although the 741 operational amplifier comes in several packages, the 8-pin "mini-DIP" (sometimes called the "V" package), whose pin connections are in Fig. 1-4, is preferred for active filters.

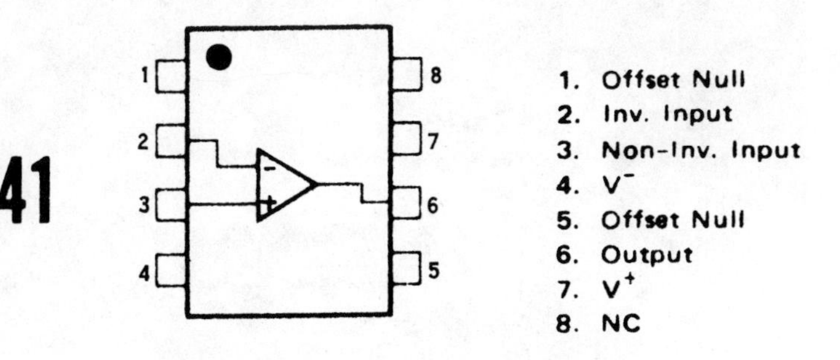

Fig. 1-4. Pinout diagram of 741 op amp.

- *The type-558 Dual Operational Amplifier.* This device (the 5558, 1458, and 1558 are other designations used) consists of a pair of independent high performance amplifiers in a single 8-pin mini-DIP package (Fig. 1-5), and sells for about 60 cents. This chip saves a lot of space when two or more operational amplifiers are required.

5558

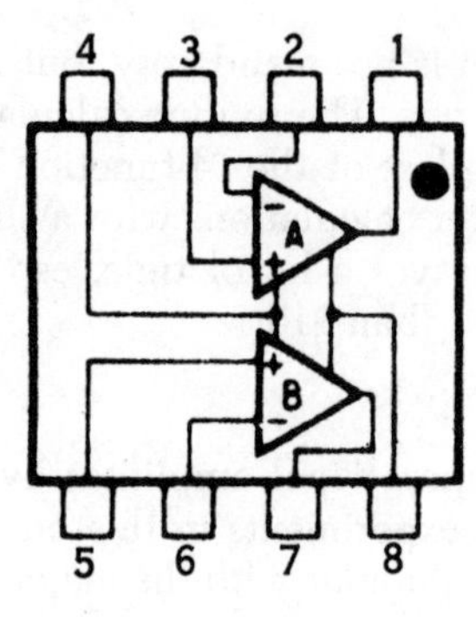

1. Output A
2. Inverting Input A
3. Noninverting Input A
4. V^-
5. Noninverting Input B
6. Inverting Input B
7. Output B
8. V^+

Fig. 1-5. Pinout diagram of 5558 op amp.

AN INTRODUCTION TO THE EXPERIMENTS

The following experiments in this chapter are designed to demonstrate the operation of the LR-31 Outboard and its use with an oscilloscope. If you have not already done so, please read the "Rules for Setting Up Experiments" section earlier in this chapter. If you are going to use a frequency source different than the LR-31 Outboard, you can omit these experiments. However, it will be helpful if you read through the experiments. Fig. 1-6 shows the types of hardware that you will generally need for the experiments in this book.

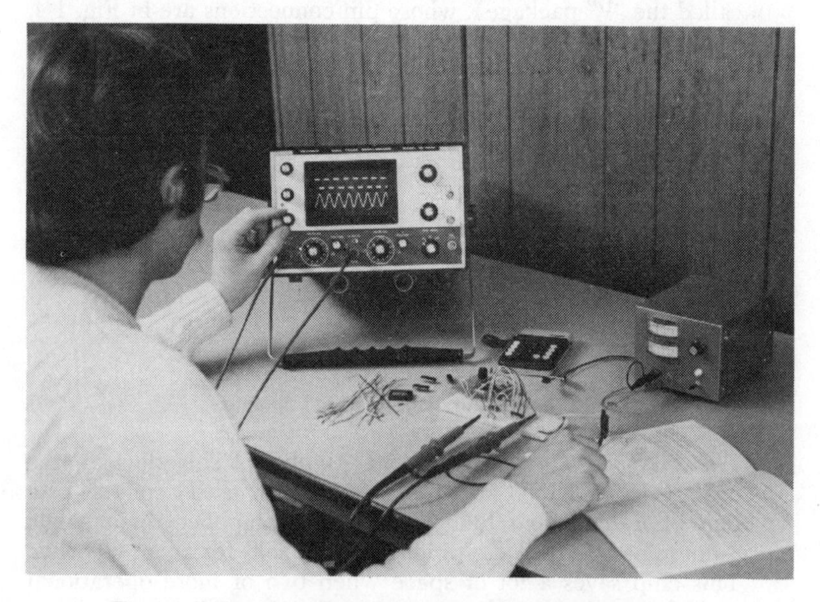

Fig. 1-6. Types of hardware generally needed for experiments.

EXPERIMENT NO. 1

Purpose

The purpose of this experiment is to demonstrate the use of the LR-31 Outboard with an oscilloscope.

Step 1

Carefully study the top view of the LR-31 Outboard, shown in Fig. 1-7 mounted on one edge of the SK-10 socket. Note that there

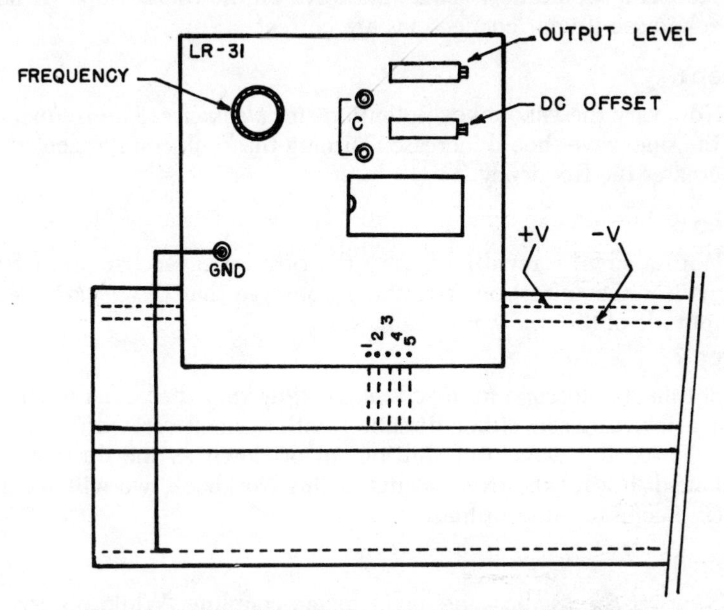

Fig. 1-7. Outboard mounted on an edge of SK-10 socket.

are two "trimpots," or multi-turn potentiometers. One is for adjusting the dc offset level, while the other is for adjusting the output level. Also, another potentiometer (single turn) is used for setting the frequency.

Step 2

Insert a .1-μF capacitor into the socket pins on the Outboard. Then connect the two pairs of alligator clips to two "lantern" batteries, or any other dual polarity supply between 6 and 12 volts, with *one red/black pair of leads for each battery.*

Step 3

Connect the oscilloscope *ground* lead to the GROUND terminal on the Outboard. Set the oscilloscope for *ac coupling,* a vertical sensitivity of about 0.5 volt/division (or 0.5 volt/cm), and the time base to about 1 ms/division.

Step 4

Now place a short jumper wire between terminals marked "3" and "4" on the LR-31 Outboard, making this connection on the SK-10 socket. Also, by means of a short wire, connect the oscilloscope probe to the terminal marked "5," also making this connection on the SK-10 socket. You should now see a sine wave on the oscilloscope. If not, check to see if your connections are correct.

Step 5

Now vary the FREQUENCY potentiometer *clockwise*. The frequency of the sine wave should increase. Turning the knob counterclockwise decreases the frequency.

Step 6

With a small screwdriver, vary the OUTPUT LEVEL trimpot a few turns in either direction. You should observe that the *peak-to-peak* amplitude of the sine wave changes.

Step 7

Set the oscilloscope for *dc coupling*. Now vary the DC OFFSET trimpot a few turns in either direction with a small screwdriver. You should see the waveform shifting up or down as the dc offset is changed. For all the experiments in this workbook, we will set the oscilloscope for ac coupling.

Step 8

Now set the oscilloscope again for ac coupling. While observing the oscilloscope display, remove the wire between terminals 3 and 4 from the SK-10 socket. You should now notice a triangular waveform instead of a sine wave. Consequently, *if a wire is placed between terminals 3 and 4, a sine wave will be present at terminal 5. If there is no connection, a triangular waveform will be present.*

The sine and triangular waveforms can be adjusted by the FREQUENCY, DC OFFSET, and OUTPUT LEVEL potentiometers. When using the sine-wave output, the schematic symbol (Fig. 1-8) will be used,

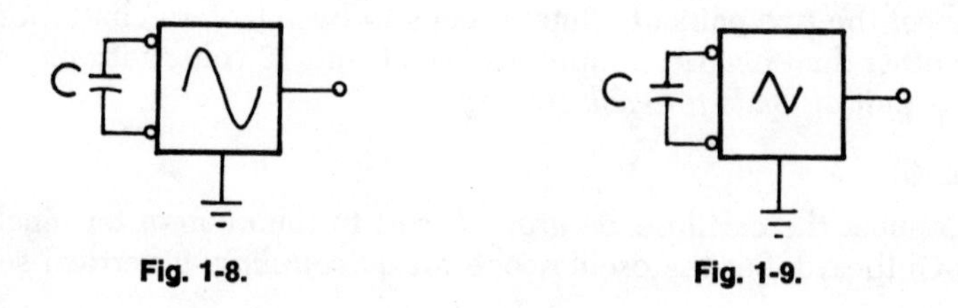

Fig. 1-8. **Fig. 1-9.**

keeping in mind that there is a connection between terminals 3 and 4.

The capacitor marked C, is the external capacitor inserted into the socket pins on the OUTBOARD. The schematic symbol in Fig. 1-9. will be used for triangular waveforms.

Step 9

Now connect a wire from terminal "1" to the + supply voltage on the SK-10 socket, which is the outermost row of pins, as shown in Fig. 1-7. Now transfer the wire from terminal 5 to terminal 2, as made on the SK-10 socket. You should now see a square-wave on the oscilloscope's display, when the input sensitivity is set at about 2 volts/division.

Step 10

Vary the FREQUENCY control in either direction. The frequency of the square wave is also controlled by this potentiometer in the same manner as for either sine waves or for triangular waveforms. However, the DC OFFSET *and* OUTPUT LEVEL *trimpots have no effect on the square-wave output signal. The output will vary from ground potential to the + supply voltage. If 6-volt lantern batteries are used to power the LR-31 Outboard, the peak-to-peak square-wave voltage will be fixed at 6 volts.* When used with a +5 volt supply, the LR-31 Outboard is able to deliver TTL-compatible signals.

Step 11

In order to reduce the square-wave output, we will use external resistors, connected as a *voltage divider* on the SK-10 socket, shown schematically in Fig. 1-10. The output voltage, as a function of these two resistors, is:

$$V_o = \frac{R_2}{R_1 + R_2} V_i$$

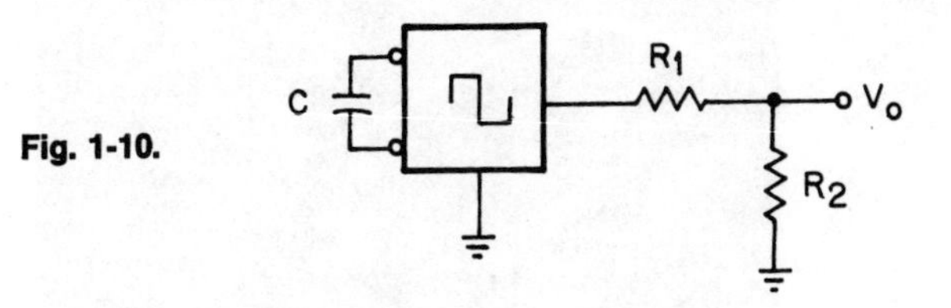

Fig. 1-10.

If, for example, we are using 6 volt batteries and $R_1 = 100\ k\Omega$ and $R_2 = 1\ k\Omega$, then the square wave will have a peak-to-peak output of:

$$V_o = \frac{1}{100 + 1}\,(6\text{ volts})$$
$$= 0.059\text{ volt.}$$

EXPERIMENT NO. 2

Purpose

The purpose of this experiment is to determine the output frequency of the LR-31 Outboard using an oscilloscope.

Step 1

Insert a .1-μF capacitor into the two socket pins marked "C" on the LR-31 Outboard. Then apply power to the breadboard, and the scope probe to the sine-wave output terminal with a wire into the SK-10 socket (terminal 5).

Step 2

Set the oscilloscope for the following settings:

- Channel 1: 100 mV/division
- Time base: 1 ms/division
- AC coupling

Step 3

Connect a wire between terminals 3 and 4 on the SK-10 socket so that there is a sine-wave output signal. Now carefully adjust the OUTPUT LEVEL control for a 200-mV peak-to-peak signal (2 full vertical divisions).

Step 4

Now carefully adjust the FREQUENCY potentiometer so that one full cycle occupies the oscilloscope screen (1 full cycle/10 horizontal divisions) as shown in Fig. 1-11. What is the output frequency?

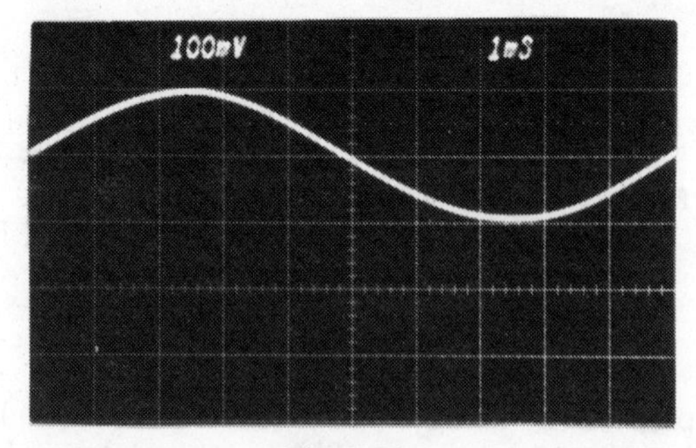

Fig. 1-11.

Your answer should have been 100 Hz. Since the total cycle is 10 divisions, the period equals:

$$T = (10 \text{ divisions})(1 \text{ ms/division})$$
$$T = 10 \text{ ms}$$

so that the output frequency is then:

$$f = \frac{1}{T}$$

$$f = \frac{1}{10 \text{ ms}}$$

$$f = 100 \text{ Hz.}$$

Step 5

Adjust the FREQUENCY control so that there are *exactly* 2, 6, and 10 complete cycles respectively for the 10 horizontal divisions, as shown in Fig. 1-12. This corresponds to frequencies of 200 Hz, 600 Hz, and 1 kHz respectively.

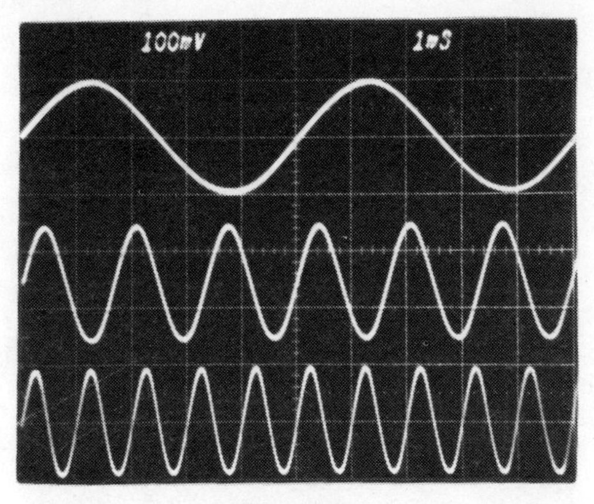

Fig. 1-12.

Step 6

Now, without disturbing the FREQUENCY control, change the scope time base to 0.1 ms/division. How many cycles do you see now?

You should see exactly 1 cycle, since the frequency is now 1 kHz. For the experiments in this book, we will be using frequencies from 100 Hz to about 10 kHz, with a .1-μF capacitor inserted in the Outboard socket pins. If the need arises, frequencies higher than 10 kHz are possible by using a .01-μF capacitor.

Since there is no calibrated frequency dial on the LR-31 Outboard, we will rely either on the counting of the number of complete cycles per 10 horizontal divisions, or determining the frequency from the period of the signal, illustrated in Step 5. As it turns out, these two methods are quite accurate for our purpose. When compared against a $1,000 frequency counter that had just been calibrated, the differences were less than 2%.

CHAPTER 2

The Operational Amplifier, The Basic Building Block

INTRODUCTION

The operational amplifier, or op amp, is the basic building block for most of the active filters now in use. For this reason, we will first begin our study of active filters by learning how op amps are used.

OBJECTIVES

At the completion of this chapter, you will be able to do the following:

- Explain the differences in the operation of the inverting and noninverting inputs
- Design and wire-up the following op-amp circuits:
  - inverting amplifier
  - noninverting amplifier
  - voltage follower
  - summing amplifier
  - integrator
- Recognize the limitations of op amps

In this chapter we are primarily concerned with how to use the op amp, concentrating only on a few basic principles and design equations. As an in-depth discussion of op amps is beyond the scope of this book, the following references are excellent to fill in the gaps:

1. Jung, W. G. *The IC Op-Amp Cookbook*. Howard W. Sams & Company, Inc., Indianapolis, 1974.

2. Tobey, G. E., et al. *Operational Amplifiers, Design and Applications.* McGraw-Hill, New York, 1971.

OP-AMP BASICS

The term "operational amplifier" originated in the field of analog computers, since these types of amplifiers were used to achieve mathematical operations such as addition and integration. Although op amps are now used for a wide range of applications, the original term is still used.

The op amp is basically a high voltage gain dc differential amplifier having ideal characteristics of:

- infinite bandwidth
- infinite input impedance
- zero output impedance

As shown in Fig. 2-1A, the op amp has a *noninverting* or + input, an *inverting* or − input, and a single output. In addition, the op amp is normally powered by a dual-polarity power supply in the range of ±5 to ±15 volts. That is, one supply is +5 to +15 volts *with respect*

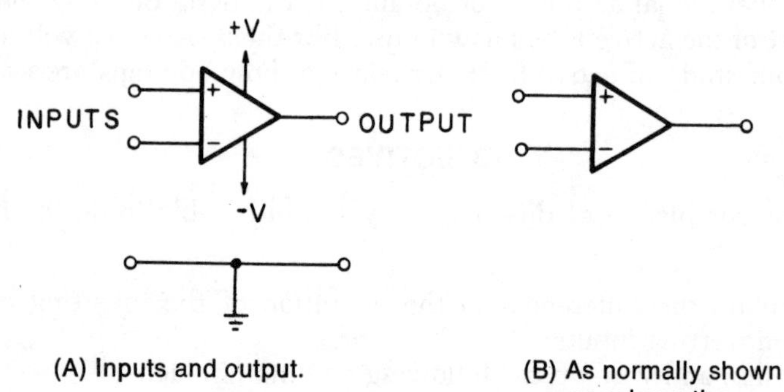

(A) Inputs and output.

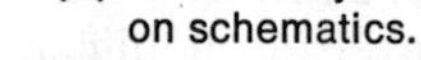

(B) As normally shown on schematics.

Fig. 2-1. Operational amplifier.

to ground, and another supply of −5 to −15 volts also *with respect to ground.* Although the op amp can be biased to operate from a single positive supply, these methods will not be discussed. In all further circuits, it will be implied that *the op amp is connected to a bipolar supply, although this connection is not specifically indicated on the schematic diagram,* as shown in Fig. 2-1B. A simple bipolar supply which you can use for your experiments is shown in Fig. 2-2, using a pair of 6-volt "lantern" batteries.

As mentioned earlier, the op amp has two inputs. The functional difference between these two inputs can be explained as follows:

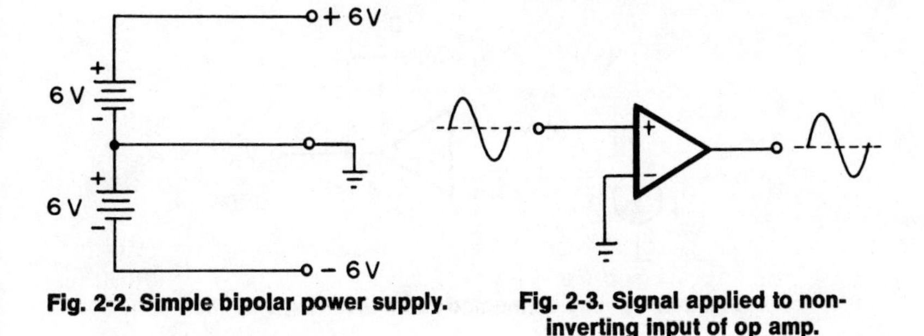

Fig. 2-2. Simple bipolar power supply.

Fig. 2-3. Signal applied to noninverting input of op amp.

- If a signal is applied to the *noninverting* (+) input of an op amp, the output will be *in phase* with the input. That is, when the input signal goes *positive,* the output also goes positive, as shown in Fig. 2-3.
- If a signal is applied to the *inverting* (−) input of an op amp, the output will be *out of phase by 180°*, or one-half cycle. That is, when the input signal goes positive, the output goes *negative*, or is *inverted with respect to the input,* as shown in Fig. 2-4.

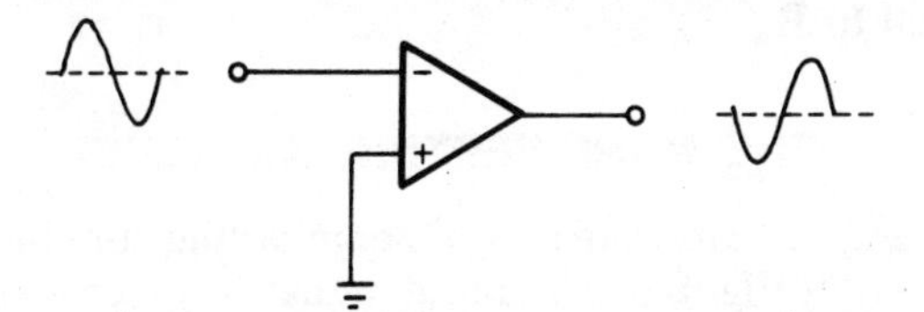

Fig. 2-4. Signal applied to inverting input of op amp.

As we shall see in this chapter, op-amp circuits used in active filters always use some form of external feedback to divert part of the output signal back to the input of the op amp, thus stabilizing the op-amp's characteristics, not to mention making the design equations simpler.

THE INVERTING AMPLIFIER

The op amp is connected as an inverting amplifier using the circuit of Fig. 2-5. R_A is called the *input element* and R_B is called the *feedback element*. For this circuit, both elements are resistors. The input is applied to the *inverting* input via R_A and the noninverting input is grounded. R_B allows a fraction of the output voltage (V_o) to be "fed back" to the inverting input. In terms of R_A and R_B, the output voltage is

$$V_o = -\frac{R_B}{R_A} V_i \qquad \text{(Eq. 2-1)}$$

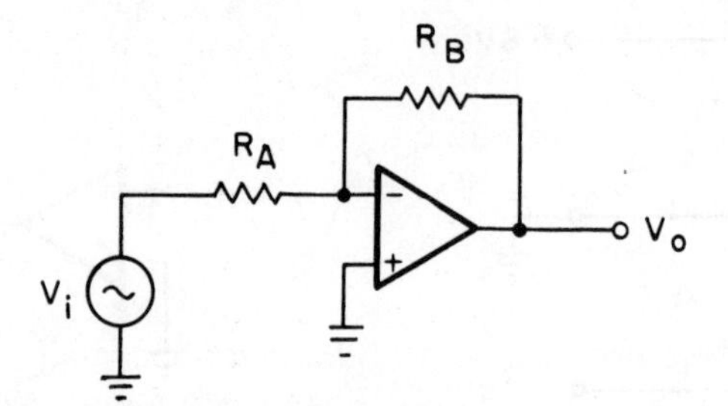

Fig. 2-5. Op amp connected as an inverting amplifier.

Consequently, the *voltage gain, or the ratio of the output voltage to the input voltage* is

$$\text{voltage gain} = \frac{V_o}{V_i} = -\frac{R_B}{R_A} \qquad \text{(Eq. 2-2)}$$

and is only dependent on *the ratio of the feedback resistance* (R_B) *to the input resistance* (R_A). Consequently, the voltage gain can either be less than 1, equal to 1, or greater than 1. Typically R_A is at least 1 kΩ, since the input impedance of the inverting amplifier circuit is equal to R_A.

THE NONINVERTING AMPLIFIER

The op amp is connected as a noninverting amplifier using the circuit shown in Fig. 2-6. The input signal is directly applied to the

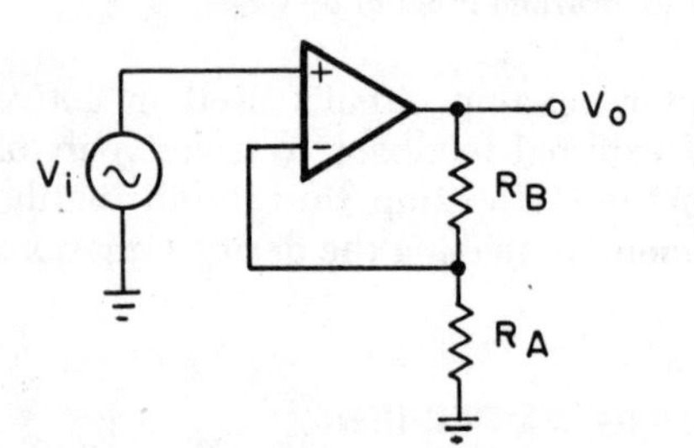

Fig. 2-6. Op amp connected as a noninverting amplifier.

noninverting input, while the input resistor is grounded. In terms of R_A and R_B, the output voltage is

$$V_o = \left(1 + \frac{R_B}{R_A}\right) V_i \qquad \text{(Eq. 2-3)}$$

so that the voltage gain is

$$\text{voltage gain} = \frac{V_o}{V_i} = 1 + \frac{R_B}{R_A} \qquad \text{(Eq. 2-4)}$$

It should be noted that, unlike the inverting amplifier, the voltage gain will *always be greater than 1 (unity)*.

THE VOLTAGE FOLLOWER

The voltage follower, shown in Fig. 2-7, is sometimes called a *buffer*, and has the same function as an emitter follower or cathode follower. It therefore has a very high input impedance (greater than 100 kΩ) and a very low output impedance (less than 75Ω). The

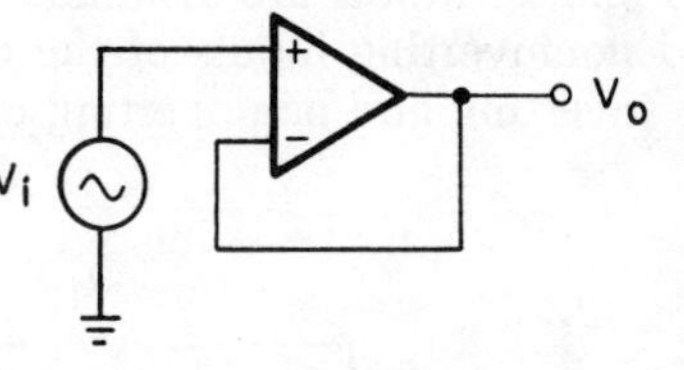

Fig. 2-7. A voltage follower.

voltage follower is similar to the noninverting amplifier except that $R_A = \infty$ and $R_B = 0$ in Fig. 2-6, so that the voltage gain is always equal to 1. The output signal is then identical to the input signal, or the output *follows* the input. Its primary function is to *buffer* or *isolate* the load (the output) from the source (the input).

THE SUMMING AMPLIFIER

We can add two or more independent input signals by using the summing amplifier circuit of Fig. 2-8. This circuit is similar to the inverting amplifier, except we now have two inputs V_1 and V_2. The

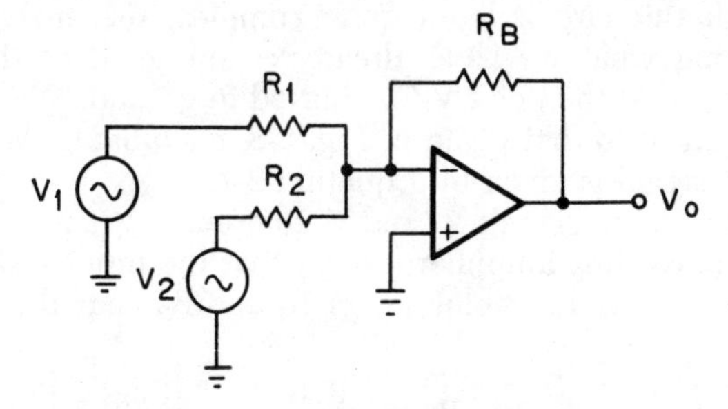

Fig. 2-8. A summing amplifier.

voltage gain for each input is determined by the ratio of the feedback resistor R_B and the corresponding input resistor:

$$\text{voltage gain}_1 = \frac{V_o}{V_1} = -\frac{R_B}{R_1} \qquad \text{(Eq. 2-5)}$$

$$\text{voltage gain}_2 = \frac{V_o}{V_2} = -\frac{R_B}{R_2} \quad \text{(Eq. 2-6)}$$

so that the output voltage is

$$V_o = -\left(\frac{R_B}{R_1}V_1 + \frac{R_B}{R_2}V_2\right) \quad \text{(Eq. 2-7)}$$

As with the circuit of Fig. 2-5, the input impedance of each input is simply the value of the corresponding input resistor.

A more useful circuit, as will be used in a later chapter, is shown in Fig. 2-9. Inputs are simultaneously applied to both the inverting and noninverting inputs of the op amp, which is a combination of the inverting and noninverting circuits already discussed.

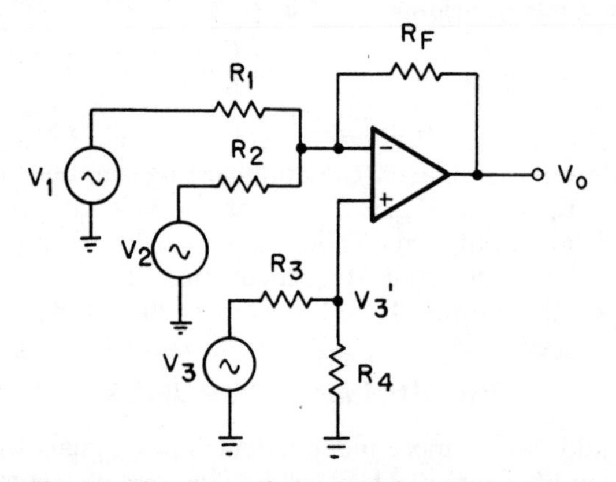

Fig. 2-9. Signals applied to both inputs of op amp.

Although this circuit looks quite complex, the analysis is quite simple, using what we have already encountered in this chapter. First assume that the point V_3' is shorted to ground. We now have a circuit identical to that given in Fig. 2-8, *a summing amplifier*, and the output signal is given by Equation 2-7.

Next, short input signals V_1 and V_2 to ground. We now have essentially a noninverting amplifier, except that the input resistor is now equal to the parallel combination of R_1 and R_2 (equal to R_A in Fig. 2-6), or

$$R_A = \frac{R_1 R_2}{R_1 + R_2} \quad \text{(Eq. 2-8)}$$

The actual noninverting input voltage that the op amp sees is V_3' which is related to V_3 by the voltage divider equation,

$$V_3' = \frac{R_4}{R_3 + R_4} V_3 \quad \text{(Eq. 2-9)}$$

The noninverting output voltage, in terms of Equation 2-3, now is

$$V_o = \left(1 + \frac{R_F}{R_A}\right) V_3' \qquad \text{(Eq. 2-10)}$$

Substituting Equations 2-8 and 2-9 into Equation 2-10, we find that

$$V_o = \left(\frac{R_1R_2 + R_FR_1 + R_FR_2}{R_1R_2}\right)\left(\frac{R_4}{R_3 + R_4}\right) V_3 \qquad \text{(Eq. 2-11)}$$

and the output voltage, as a function of the three inputs V_1, V_2, and V_3 is

$$V_o = -\frac{R_F}{R_1} V_1 - \frac{R_F}{R_2} V_2 + \left(\frac{R_1R_2 + R_FR_1 + R_FR_2}{R_1R_2}\right)\left(\frac{R_4}{R_3 + R_4}\right) V_3 \qquad \text{(Eq. 2-12)}$$

The first two terms are the *inverted* outputs, and the last term is the *noninverted* output. From Equation 2-12, if we wish to change the voltage gain of either of the two inverting input signals, we cannot do so without affecting the voltage gain of the noninverting input signal!

THE INTEGRATOR

By changing the feedback resistor of the inverting amplifier circuit of Fig. 2-5 to a capacitor, an op-amp integrator is formed, as shown

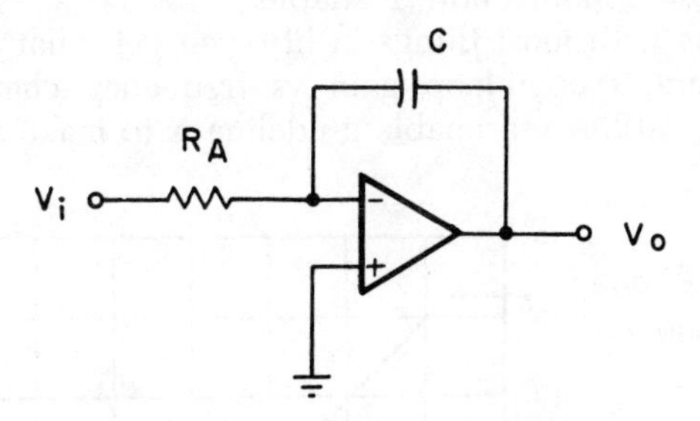

Fig. 2-10. An op-amp integrator.

in Fig. 2-10. The input signal is *integrated* and represents the *"area under the curve."* The output voltage is given by the relation

$$V_o = -\frac{1}{R_AC}\int_0^t V_i \, dt \qquad \text{(Eq. 2-13)}$$

The term $1/R_AC$ should be matched for the minimum input frequency expected, so that

$$R_A C = \frac{1}{2\pi f_{minimum}} \quad \text{(Eq. 2-14)}$$

Since the integrator will also act on any resultant offset voltage output created, due to the op-amp's bias current offset, a resistor equal to R_A is frequently placed between the noninverting input and ground to minimize this offset, as shown in Fig. 2-11.

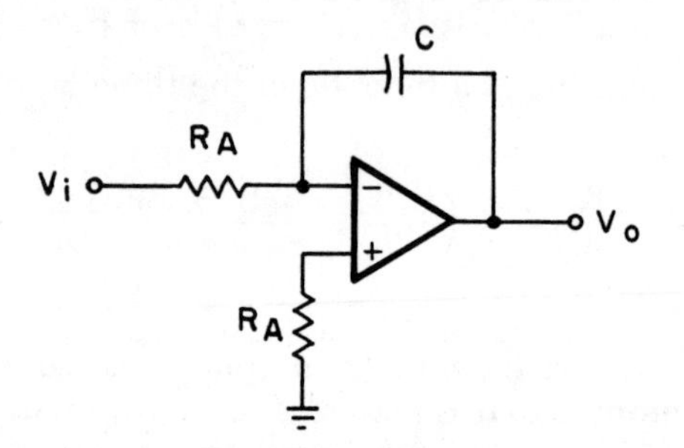

Fig. 2-11. Resistor used to minimize offset error.

SELECTING THE RIGHT OP AMP

All of the circuits presented in this chapter will work with *all* op amps. However, when used in building active filters, certain types may be preferred over others. Perhaps the most popular op amp is the type 741, as explained briefly in Chapter 1, and is typical of the general purpose op amps now available.

However, as with most things in life, you get what you pay for. The 741 op-amp's open-loop gain vs frequency characteristic is shown in Fig. 2-12. A reasonable guideline is to make sure that the

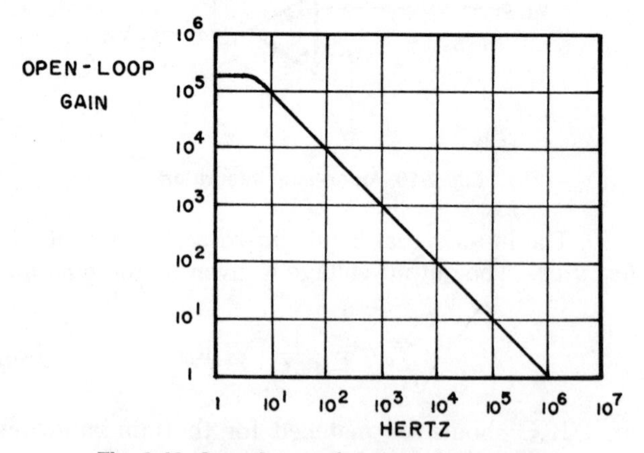

Fig. 2-12. Open-loop gain versus frequency.

op-amp's open-loop gain is a minimum of 20 times the required voltage gain of the circuit at the highest frequency of interest. For example, if we want to use a 741 op amp as an amplifier at 10 kHz, we see from the graph in Fig. 2-12 that the 741's open-loop voltage gain is approximately 100. Consequently, for the 741 op amp to perform properly, the circuit's voltage gain (e.g., a noninverting amplifier) should be no greater than 5.

Another factor affecting the performance of op amps is the *slew rate*, which affects the op-amp's *high-frequency response*. The slew rate is a *measure of how fast a large-signal can change, and is dimensionally expressed in terms of volts/μs*. For the 741 op amp, the slew rate is typically 0.5 V/μs. If the input signal, for example a sine wave, is 1 volt peak-to-peak, it will take 1 volt/0.5 volt/μs, or 2 μs to change 1 volt. Since there are two 1-volt swings for each cycle (Fig. 2-13), a total of 4 μs is the highest period that the 741 op-amp can tolerate for a 1 volt peak-to-peak signal. Therefore, the highest frequency this op amp can tolerate of a 1 volt peak-to-peak signal is

$$f_{max} = \frac{\text{slew rate}}{2 \times \text{input voltage}} \qquad \text{(Eq. 2-15)}$$

$$= \frac{0.5\ \text{V}/\mu\text{s}}{(2)(1\text{V})}$$

$$= 250\ \text{kHz}$$

Consequently, *as the input peak-to-peak voltage increases, the maximum possible operating frequency of the op amp decreases.*

However, the 741 family of op amps is fine if we are contented with the cited frequency response and slew rate. However, there are other op amps, such as the type 318, which have improved responses. The slew rate for the 318 op amp is typically 70 V/μs, consequently the maximum frequency for a 1 volt peak-to-peak signal will be 35 MHz! This op amp sells for about three times as much as the 741 op amp. For the experiments throughout this book, we will use the 741

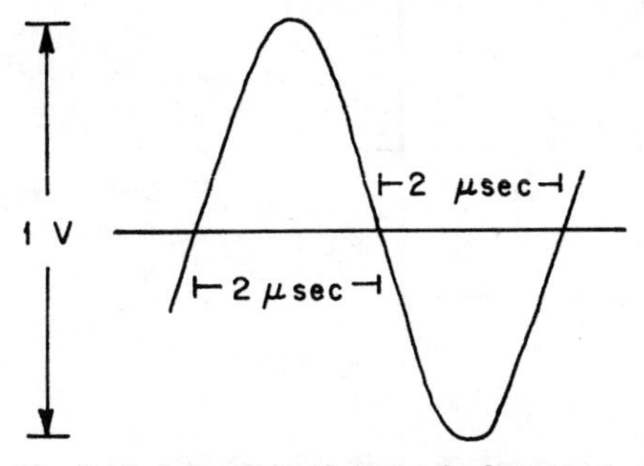

Fig. 2-13. A 1-volt peak-to-peak sine wave.

op amp, because if you accidentally damage it, you won't feel too bad! For those of you who want to go first class, use the 318 op amp.

AN INTRODUCTION TO THE EXPERIMENTS

The following experiments are designed to demonstrate the design and operation of the basic operational amplifier circuits presented in this chapter. Once you are convinced of the performance of these basic circuits, the design of active filter circuits will be simplified.

The experiments that you will perform can be summarized as follows:

Experiment No.	*Purpose*
1	Demonstrates the operation of the inverting amplifier.
2	Demonstrates the operation of the noninverting amplifier.
3	Demonstrates the operation of a voltage follower.
4	Demonstrates the operation of a summing amplifier.
5	Demonstrates the operation of an integrator.

EXPERIMENT NO. 1

Purpose

The purpose of this experiment is to demonstrate the operation of the inverting amplifier, using a type 741 op amp.

Pin Configuration of 741 Op Amp (Fig. 2-14)

Note: we are using the 8-pin mini-DIP chip!

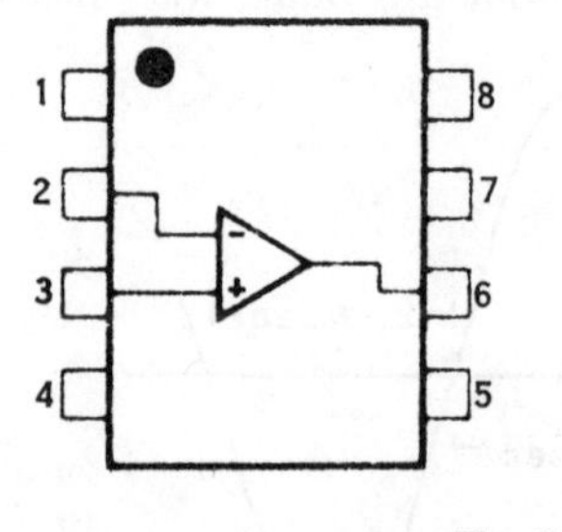

1. Offset Null
2. Inv. Input
3. Non-Inv. Input
4. V^-
5. Offset Null
6. Output
7. V^+
8. NC

Fig. 2-14.

Schematic Diagram of Circuit (Fig. 2-15)

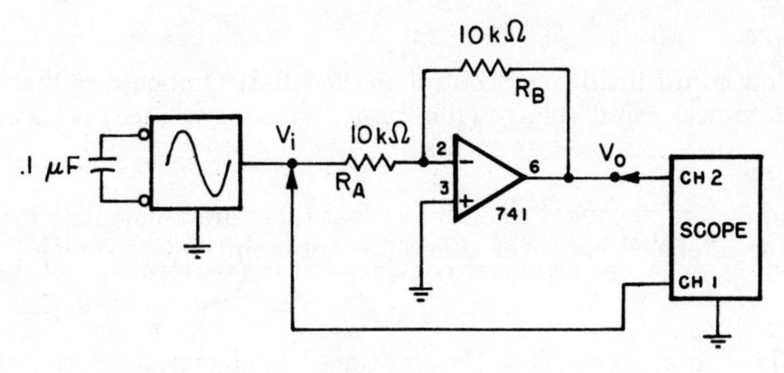

Fig. 2-15.

Design Basics

$$\text{Gain} = \frac{V_o}{V_i} = -\frac{R_B}{R_A}$$

Step 1

If you haven't worked with op amps before, the +V and −V pin connections are *usually omitted from schematic diagrams, as the connections are implied!*

Remember: *always connect pin 7 to +V and pin 4 to −V, using the 8-pin type 741 op-amp package.*

Step 2

Check your wired circuit, making sure that it is correct. Don't forget the +V and −V connections.

Step 3

Set the oscilloscope for the following settings:

- Channels 1 & 2: 0.1 volt/division
- Time base: 1 ms/division
- AC coupling

Apply power to the breadboard and observe the two traces on the scope screen.

Note: Since we will be concerned with both the input and output signals, we will adopt the convention that the input signal is Channel 1, and the output signal is Channel 2. When viewing both signals simultaneously on a dual-trace oscilloscope, *position the input signal so that it is above the output signal.*

Step 4

Now adjust the OUTPUT control on the LR-31 Outboard so that the peak-to-peak input voltage (the "upper" sine-wave trace) is 0.2 volt.

Step 5

Adjust the FREQUENCY control so that there are 2 complete cycles on the screen. What is the difference between the two signals?

The output signal (the "lower" trace) is of opposite form, or is *inverted*, compared to the input signal, as shown in Fig. 2-16. The output is then said to be inverted, or 180° *out-of-phase* with the input, since the positive peak of the output signal occurs when the input peak is negative.

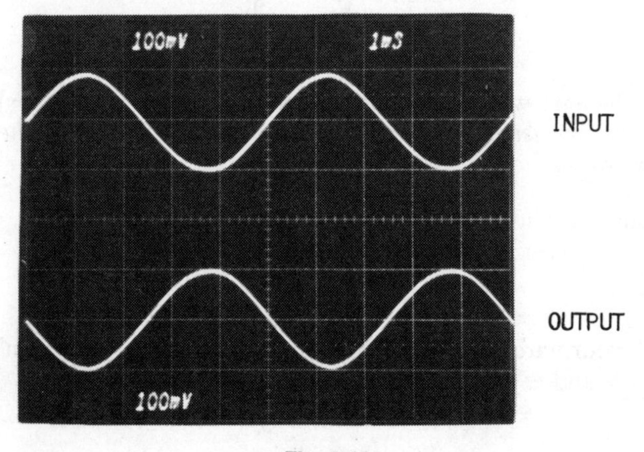

Fig. 2-16.

Step 6

What is the peak-to-peak output voltage?

You should have answered, about 200 mV. Using the voltage gain as *the ratio of the output voltage to the input voltage*, how does this compare with the equation given in the "Design Basics" section?

The voltage gain is −1.0 (i.e., 200 mV/200 mV). Also by the equation,

$$\text{voltage gain} = -\frac{R_B}{R_A}$$

$$= -\frac{10\ k\Omega}{10\ k\Omega}$$
$$= -1.0$$

The *minus sign* indicates that the output is inverted with respect to the input.

Step 7

Keeping the input level constant at 200 mV peak-to-peak, change R_B and complete the following table. Do your experimental results compare with the design equation?

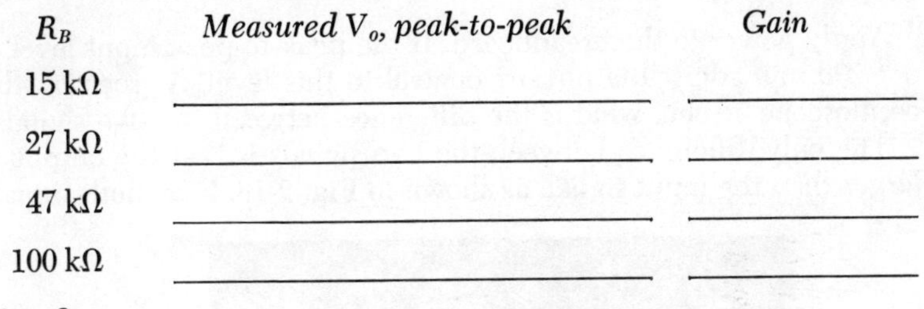

R_B	*Measured V_o, peak-to-peak*	*Gain*
15 kΩ		
27 kΩ		
47 kΩ		
100 kΩ		

Step 8

Disconnect the power from the breadboard and proceed to the next experiment. Try not to disturb the voltage and frequency settings of the function generator, as they will remain the same.

EXPERIMENT NO. 2

Purpose

The purpose of this experiment is to demonstrate the operation of the noninverting amplifier, using a type 741 op amp.

Schematic Diagram of Circuit (Fig. 2-17)

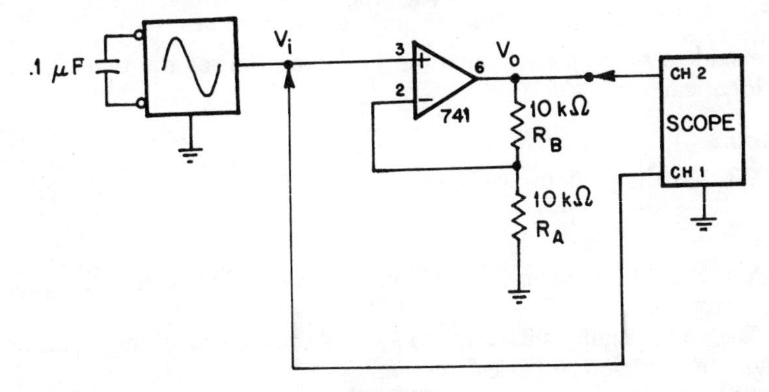

Fig. 2-17.

Design Basics

$$\text{Gain} = \frac{V_o}{V_i} = 1 + \frac{R_B}{R_A}$$

Step 1

Set the oscilloscope for the following settings:

- Channels 1 & 2: 0.1 volt/division
- Time base: 1 ms/division
- AC coupling

Step 2

Apply power to the breadboard. If the peak-to-peak input level is not 200 mv, adjust the OUTPUT control to this level. As seen on the oscilloscope screen, what is the difference between the two signals?

The only difference between the two signals is that the output is *larger* than the input signal, as shown in Fig. 2-18. Now both signals

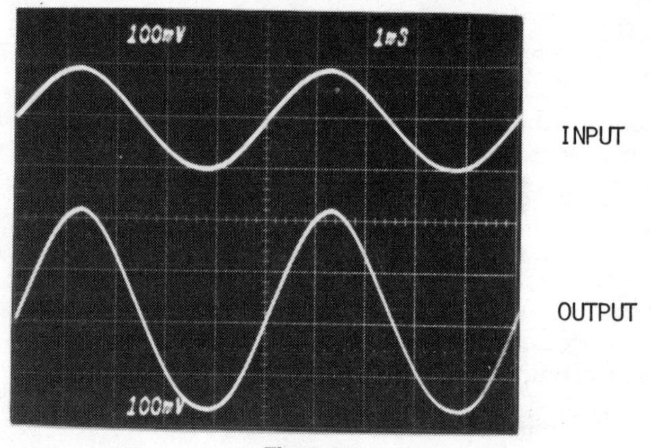

Fig. 2-18.

are said to be *in phase*, since the time variation of the two signals is identical.

Step 3

What is the peak-to-peak output voltage?

You should have measured about 400 mV. What then is the voltage gain?

The voltage gain is 2.0. How does this compare with the equation given in the "design basics" section?

By the equation,

$$\text{gain} = 1 + \frac{R_B}{R_A}$$

$$= 1 + \frac{(10\ k\Omega)}{(10\ k\Omega)}$$

$$= 2.0$$

Step 4

Keeping the input level constant at 200 mV, change R_B and complete the following table. Do your experimental results agree with the design equation?

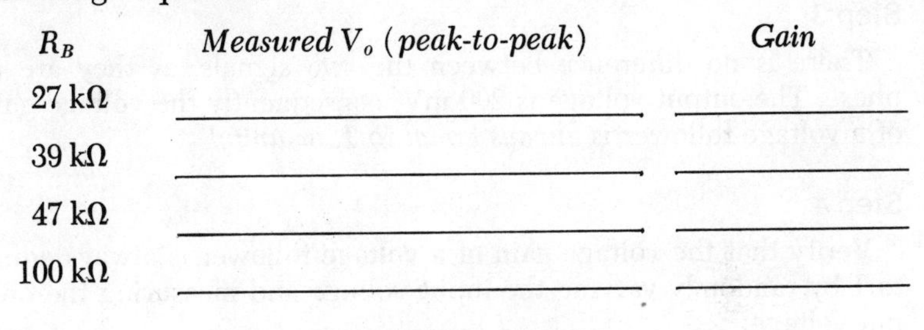

R_B	*Measured V_o (peak-to-peak)*	*Gain*
27 kΩ		
39 kΩ		
47 kΩ		
100 kΩ		

EXPERIMENT NO. 3

Purpose

The purpose of this experiment is to demonstrate the operation of a voltage follower.

Schematic Diagram of Circuit (Fig. 2-19)

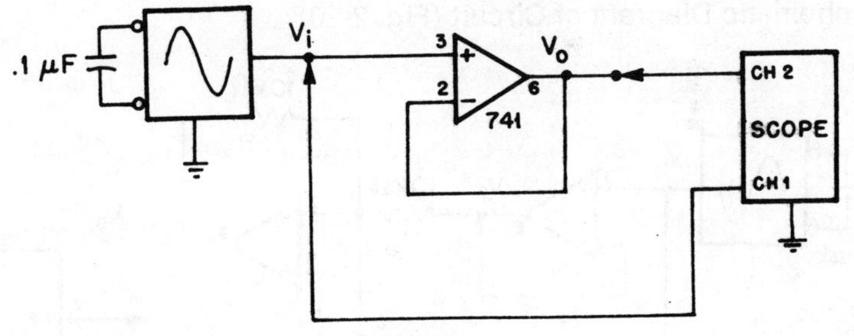

Fig. 2-19.

Design Basics

$$\text{Gain} = \frac{V_o}{V_i} = 1$$

Step 1

Set the oscilloscope for the following settings:

- Channels 1 & 2: 0.1 volt/division
- Time base: 1 ms/division
- AC coupling

Step 2

Apply power to the breadboard and adjust the peak-to-peak input voltage at 200 mV, and so that there are at least 2 complete cycles on the oscilloscope screen. What is the difference between the input and output signals, and what is the output voltage?

Step 3

There is no difference between the two signals, as they are in phase. The output voltage is 200 mV; consequently the voltage gain of a voltage follower is *always equal to 1, or unity!*

Step 4

Verify that the voltage gain of a voltage follower is always equal to 1 by randomly varying the input voltage and measuring the output voltage.

EXPERIMENT NO. 4

Purpose

The purpose of this experiment is to demonstrate the operation of the summing amplifier.

Schematic Diagram of Circuit (Fig. 2-20)

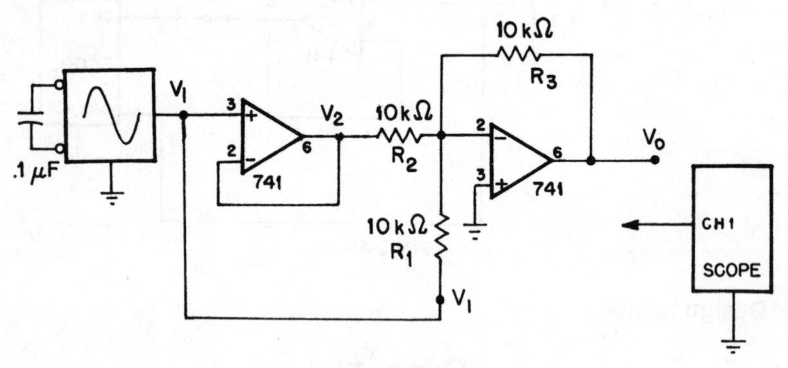

Fig. 2-20.

Design Basics

$$V_o = V_1 + V_2$$

$$= -\frac{R_3}{R_1}V_1 - \frac{R_3}{R_2}V_2$$

where $R_1 = R_2$

Step 1

Set the oscilloscope for the following settings:

- Channel 1: 1 volt/division
- Time base: 1 ms/division
- AC coupling

Step 2

Apply power to the breadboard and adjust the peak-to-peak output voltage of the LR-31 Signal Generator (V_1) to 1 volt, and the frequency so that there are 3 full cycles on the scope screen (300 Hz).

Step 3

Measure the voltage at the output of the 1st op amp (V_2). What is it?

You should have measured a peak-to-peak voltage of 1 volt, since this portion of the circuit is just a voltage follower, whose operation was described in the previous experiment.

Step 4

Measure the voltage at the output of the 2nd op amp (V_o). What is it?

You should have measured a peak-to-peak voltage of about 2.0 volts. This second op amp is the *summing amplifier*, adding the two input voltages V_1 (1 volt) and V_2 (also 1 volt). From the second design equation, the output voltage of the summing amplifier, shown in Fig. 2-21 is:

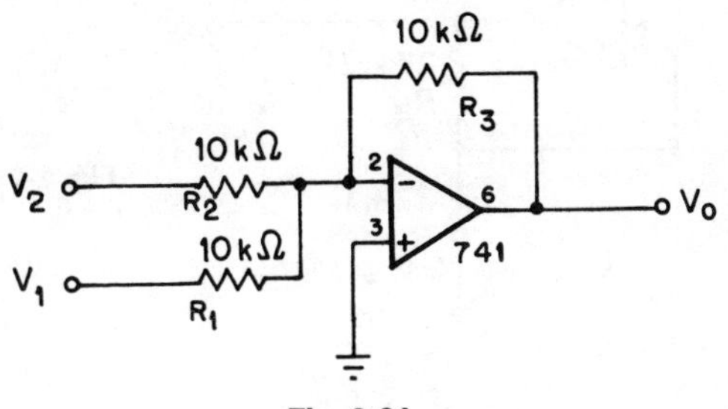

Fig. 2-21.

$$V_o = -\frac{R_3}{R_1}V_1 - \frac{R_3}{R_2}V_2$$
$$= -\frac{10\ k\Omega}{10\ k\Omega}(1\ \text{volt}) - \frac{10\ k\Omega}{10\ k\Omega}(1\ \text{volt})$$
$$= -2.0\ \text{volts}$$

The negative sign occurs because we are using the op amp as an *inverting amplifier,* so that the output is inverted with respect to the two inputs, which are in-phase (see Experiment No. 1).

If we are able to simultaneously observe V_1, V_2, and V_o on the oscilloscope, the three traces would look like that in Fig. 2-22.

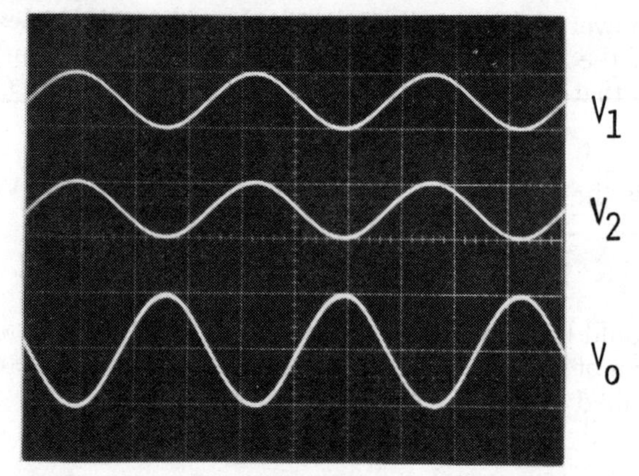

Fig. 2-22.

Step 5

So far we have only presented the simple case of adding two equal voltages. To demonstrate that the equation in Step 4 and the operation of the summing amplifier still holds for *unequal* input voltages, disconnect the power from the breadboard and rewire only the first op amp as a noninverting amplifier, as shown in Fig. 2-23. The sec-

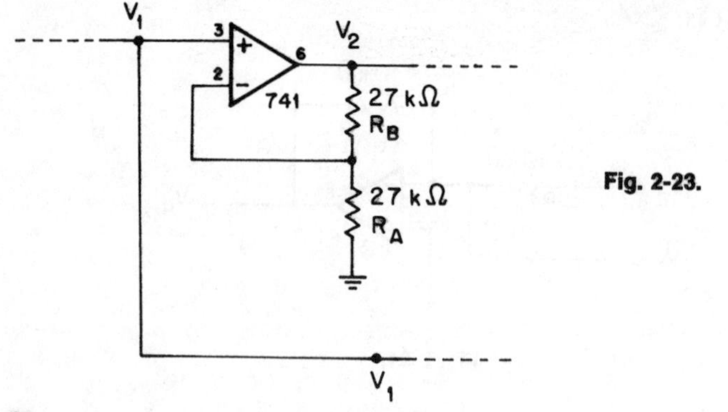

Fig. 2-23.

ond op amp remains connected as before.

Step 6

Apply power to the breadboard. What is V_2 (i.e., the output voltage of the new circuit for the 1st op amp)? Is it what you would expect?

You should have measured approximately 2 volts, since the voltage gain of this noninverting amplifier is 2.

Step 7

Now measure V_o (the output voltage of the 2nd op amp). What is it?

The peak-to-peak voltage should be approximately 3 volts. If we were again able to simultaneously observe V_1, V_2, and V_o, the traces would look like those in Fig. 2-24.

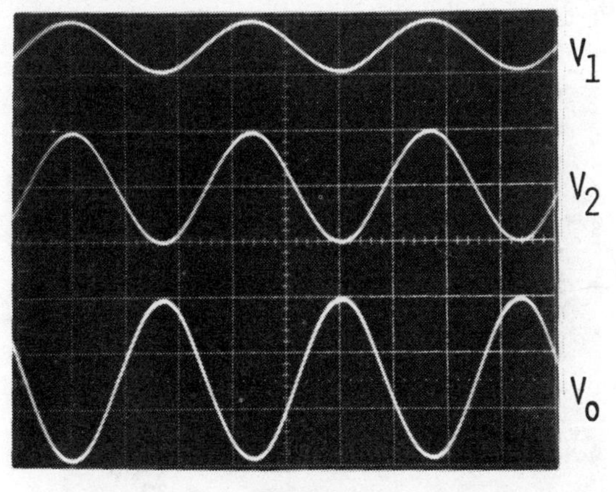

Fig. 2-24.

Step 8

Again disconnect the power and rewire the *first* op amp, as shown in Fig. 2-25.

Step 9

Apply power to the breadboard and now measure V_o (the output voltage of the second op amp). What do you get?

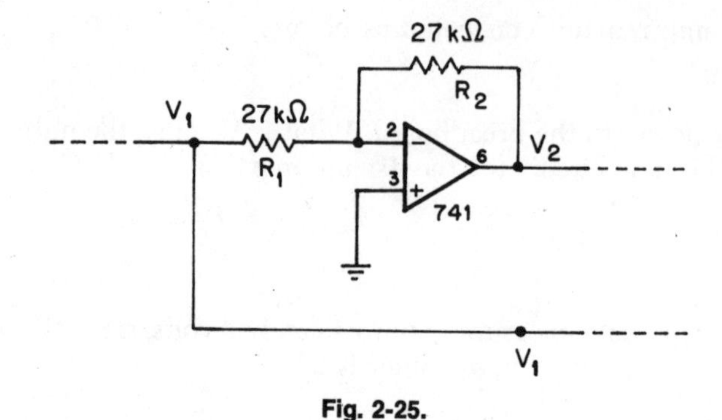

Fig. 2-25.

You should obtain *no output voltage!* Why?

You would probably think that the output voltage (V_o) would be 2 volts, since V_1 and V_2 are now each 1 volt. I have played a little trick on you. In Step 8 we were using a unity gain *inverting amplifier*, so that the output voltage (V_2) was inverted with respect to its input, V_1. When these two *equal, but out-of-phase voltages are added, they cancel each other, resulting in a net output of zero.* This can be shown by looking at V_1, V_2, and V_o simultaneously, as shown in Fig. 2-26.

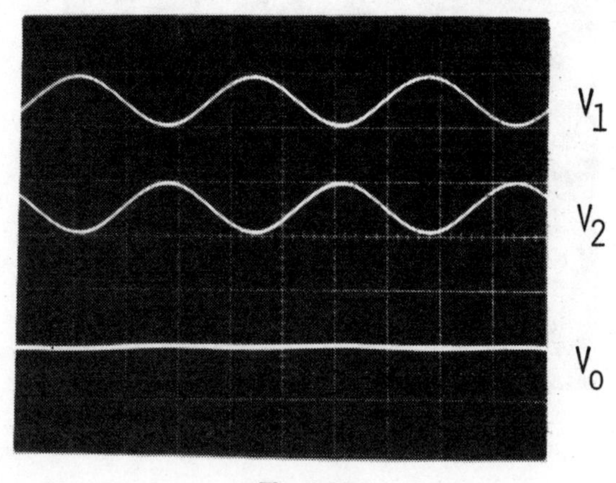

Fig. 2-26.

When V_1 goes *positive*, V_2 goes negative by an equal amount, and when they are added, the net result is zero. The same reasoning applies when V_1 goes negative. In Steps 1-7, the two input voltages were always *in-phase*. Throughout this experiment, we have added two input voltages, whether in-phase or out-of-phase.

EXPERIMENT NO. 5

Purpose

The purpose of this experiment is to demonstrate the operation of an op-amp integrator.

Schematic Diagram of Circuit (Fig. 2-27)

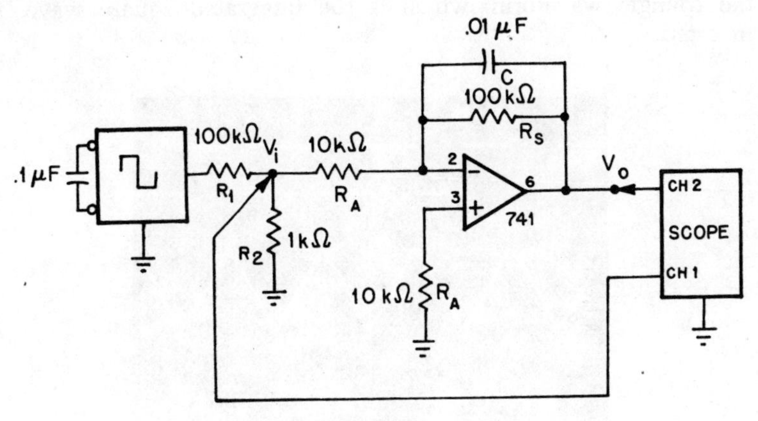

Fig. 2-27.

Design Basics

$$R_AC = \text{period of signal to be integrated}$$
$$= 1/f_1$$
$$R_S = 10\,R_A$$

$$\text{When: } f < f_1, \text{ inverting amplifier: } V_o = -\frac{R_S}{R_A} V_i$$

$$f > f_1, \text{ integrator: } V_o = -\frac{1}{R_AC}\int V_i\,dt$$

Step 1

Wire the circuit shown in the schematic diagram. If you are using a function generator other than the LR-31 Outboard, omit resistors R_1 and R_2. Set the oscilloscope for the following settings:

- Channels 1 & 2: 0.5 volt/division
- Time base: 0.5 ms/division (500 μs/division)
- AC coupling

When using the LR-31 to produce square waves, don't forget to connect terminal 1 to +V of the power supply.

Step 2

Apply power to the breadboard and adjust the frequency so that there are 5 complete cycles. If using a generator other than the LR-31, set the peak-to-peak voltage, V_i, at 0.5 volt, as shown in Fig. 2-28. The output signal (the "lower" trace) should resemble a triangle wave. Since the op amp is used in the inverting configuration (since the input is applied to the inverting input of the op amp), the triangle waveform, which is the integrated square wave, is inverted.

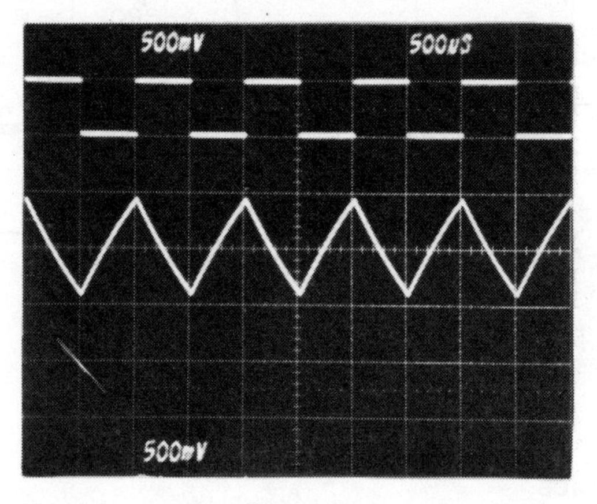

Fig. 2-28.

Step 3

Now increase the frequency by varying the FREQUENCY potentiometer. Does the output voltage of the integrator (V_o) increase or decrease? The output voltage decreases, since the impedance of the feedback capacitor is now less than the 100 kΩ resistor.

Step 4

Set the oscilloscope time base to 2 ms/division, and adjust the frequency so that there are 2 complete cycles per the 10 horizontal divisions (100 Hz). In addition, set the scope for dc coupling.

You should observe that the output does not now resemble a triangle wave, but rather a "rounded-off" square wave. In addition, the output voltage has also increased.

The difference between the circuit in this experiment and the one shown in Fig. 2-10, is the 100-kΩ resistor in the feedback loop in parallel with the .01-μF capacitor which is used to minimize effects due to the *leakage resistance* of the capacitor. Typically this *shunt resistor* (R_S) is ten times R_A.

As summarized in the "Design Basics" section, the product of R_A and C equals the period of the signal that is to be integrated, i.e. 1 kHz, or

$$\frac{1}{1\ \text{kHz}} = R_A C$$

$$1\ \text{ms} = (10\ \text{k}\Omega)(.01\ \mu\text{F})$$

Below 1 kHz, *the circuit behaves as an inverting amplifier with a gain of approximately 10.* Above 1 kHz, *the circuit acts as an integrator.* These two statements were demonstrated in Steps 3 and 4.

CHAPTER 3

Filter Basics

Before we can start designing any type of active filter, certain basic terms and concepts that will be used many times in this book must first be discussed.

OBJECTIVES

At the completion of this chapter, you will be able to do the following:

- Define and compare the general frequency responses of the following filters:
  - low-pass
  - high-pass
  - bandpass
  - band reject
- Define the following terms:

amplitude response	decibel
bandwidth	octave
cascading	order
center frequency	passband
cutoff frequency	quality, or Q
damping	rolloff
decade	stopband

THE LOW-PASS FILTER

Simply stated, *a low-pass filter allows incoming signals to be passed through with little or no attenuation up to a certain frequency; above this frequency, the filter rejects the input signal.*

To better understand the meaning of the previous statement, consider the generalized low-pass filter characteristic curve shown in Fig. 3-1. When working with filters, we are primarily interested with the relationship of the filter's output voltage compared to its input voltage at various frequencies, called the *amplitude response,* which

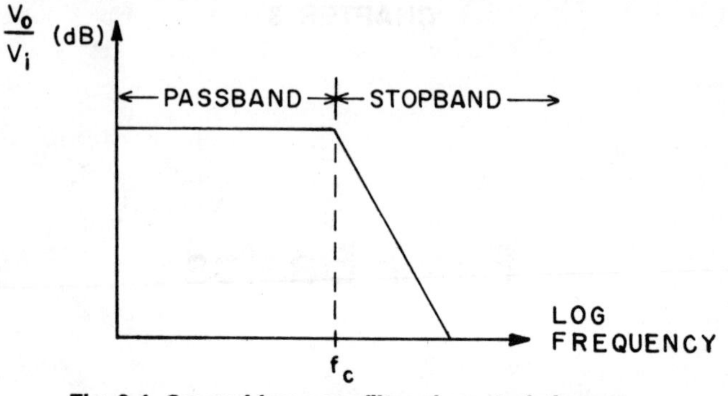

Fig. 3-1. General low-pass filter characteristic curve.

is graphically shown in Fig. 3-1. As with op amps, the filter's voltage gain, V_o/V_i, expresses this amplitude response, but as a function of frequency. For filters, the ratio V_o/V_i is more conveniently expressed in terms of *decibels,* or dB, so that by definition

$$dB = 20 \log_{10} \frac{V_o}{V_i} \qquad \text{(Eq. 3-1)}$$

If the filter's output voltage is larger than its input voltage (V_i), we naturally have gain, and the dB value will be a *positive* number. On the other hand, if the output voltage is less than the input voltage, we have a *loss,* or *attenuation,* and the dB value will be a *negative* number.

If you have a pocket calculator that has a "$\log_{10}$" function, you are in good shape. However, if you only have a simple "4-function" calculator, $\log_{10}$ of any number can be reasonably computed from the relationship*

$$\log_{10}(X) = 0.86304\left(\frac{X-1}{X+1}\right) + 0.36415\left(\frac{X-1}{X+1}\right)^3 \qquad \text{(Eq. 3-2)}$$

$$\frac{V_o}{V_i}$$

As an example, suppose $V_o/V_i = 0.707$, then using Equation 3-2,

* *Handbook of Mathematical Functions,* ed. by M. Abramowitz and I. Stegun, National Bureau of Standards Applied Math Series, 1964.

$$\log_{10}(0.707) = 0.86304\left(\frac{0.707-1}{0.707+1}\right) + 0.36415\left(\frac{0.707-1}{0.707+1}\right)^3$$
$$= 0.86304(-0.172) + 0.36415(-0.172)^3$$
$$= -0.15030$$

which is about 0.2% off from the actual value of −0.15051. For 2-place decimal accuracy, this formula is accurate.

Now back to Fig. 3-1. We see that the low-pass filter's amplitude response is essentially constant up to some frequency f_c, called *the cutoff frequency.* The range of frequency occupied by the wanted signal (in the region of constant amplitude response, or constant voltage gain) is called the *passband.* As the input signal frequency exceeds the low-pass filter's cutoff frequency, the dB amplitude response *decreases linearly as the frequency increases in a logarithmic fashion.* The frequencies above the filter cutoff frequency is called the *stopband.*

An *octave is either a doubling or halving of frequency.* For a frequency, say 1 kHz, octaves *above* 1 kHz are 2, 4, 8 kHz, etc. Going the other way, octaves *below* 1 kHz are 500, 250, 125 Hz, etc. A *decade is either a ten-fold increase or decrease in frequency.* For the 1-kHz frequency, decades above are 10, 100 kHz, etc., and decades below are 100, 10, 1 Hz, etc.

From Fig. 3-1, the dB amplitude response in the stopband decreases linearly with increasing logarithm of frequency. The rate of this decrease (or the slope of the line), called the *rolloff or falloff,* is defined by the *order of the filter.* Above the cutoff frequency, *the rolloff of a 1st order low-pass filter is −6 dB per octave.* In terms of decades, this is equivalent to *−20 dB per decade.*

A *2nd order low-pass filter* has a rolloff equal to *twice* that of a 1st order filter, or *−12 dB/octave* (*−40 dB/decade*). Consequently, a 3rd order filter has a rolloff equal to *−18 dB/octave or −60 dB/decade,* and so on for higher order low-pass filters. As graphed in Fig. 3-2, increasing the order of a low-pass filter increases the rolloff. Naturally, an infinite order filter has the best possible response.

As we shall see later on, we cannot efficiently build higher than a 2nd order filter using a single operational amplifier network. To build, for example, a 3rd order filter, we will connect in series, or *cascade,* a 1st order and a 2nd order low-pass section. For high order sections, one or more 1st and 2nd order sections are systematically cascaded, as will be explained in Chapter 7.

Up to this point we have only described the cutoff frequency as the separation point between the passband and the stopband, with Figs. 3-1 and 3-2 being only a simplistic representation of the amplitude response of low-pass filters. In reality, *the cutoff frequency of a filter is that frequency which the voltage gain drops to 0.707, or*

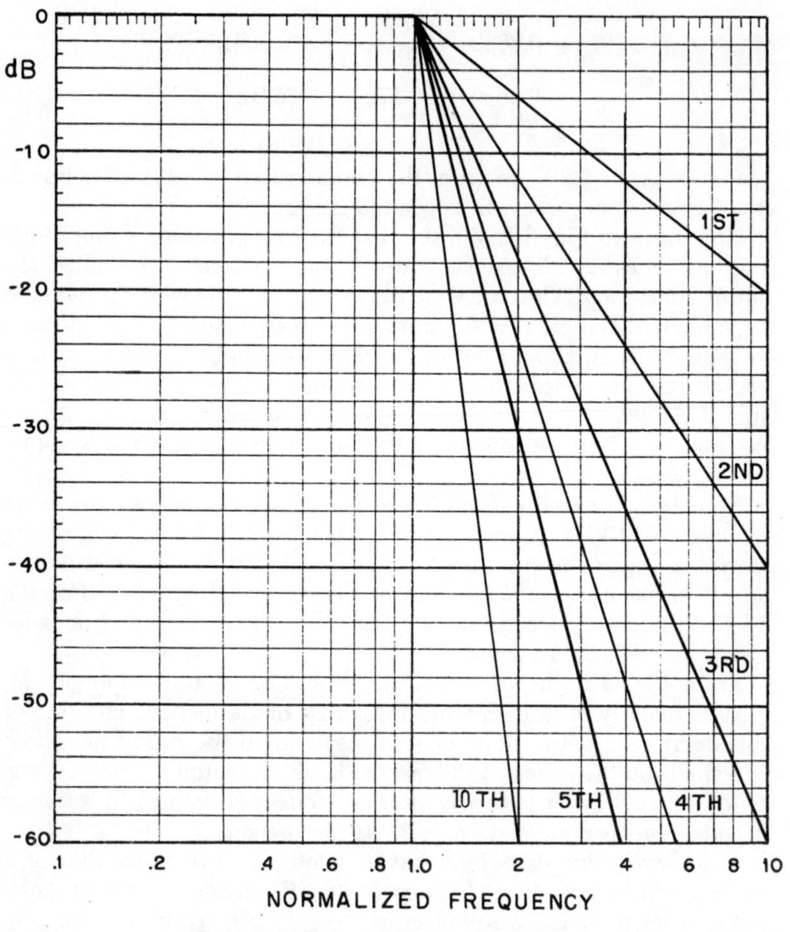

Fig. 3-2. Rolloff of different order filters.

$1/\sqrt{2}$ times the passband gain. As shown in Fig. 3-3, the voltage gain response follows a smooth line from the passband through a point that is 3 dB below the passband gain, remembering that $20 \log_{10}(0.707) = -3$ dB. For this graph, we have, for simplicity, assumed that the passband voltage gain is unity (0 dB). If on the other hand, the passband gain is 6 dB, the cutoff frequency will occur when the response falls to $6 - 3 = 3$ dB. 6 dB is equivalent to a voltage gain of 2.0, so that the cutoff frequency will be the frequency at which the voltage gain falls to 2.0 times 0.707 or 1.414, which is equivalent to 3 dB.

The graph of Fig. 3-3 illustrates a particular *shape option* that is possible, which describes the shape of the filter's amplitude response.

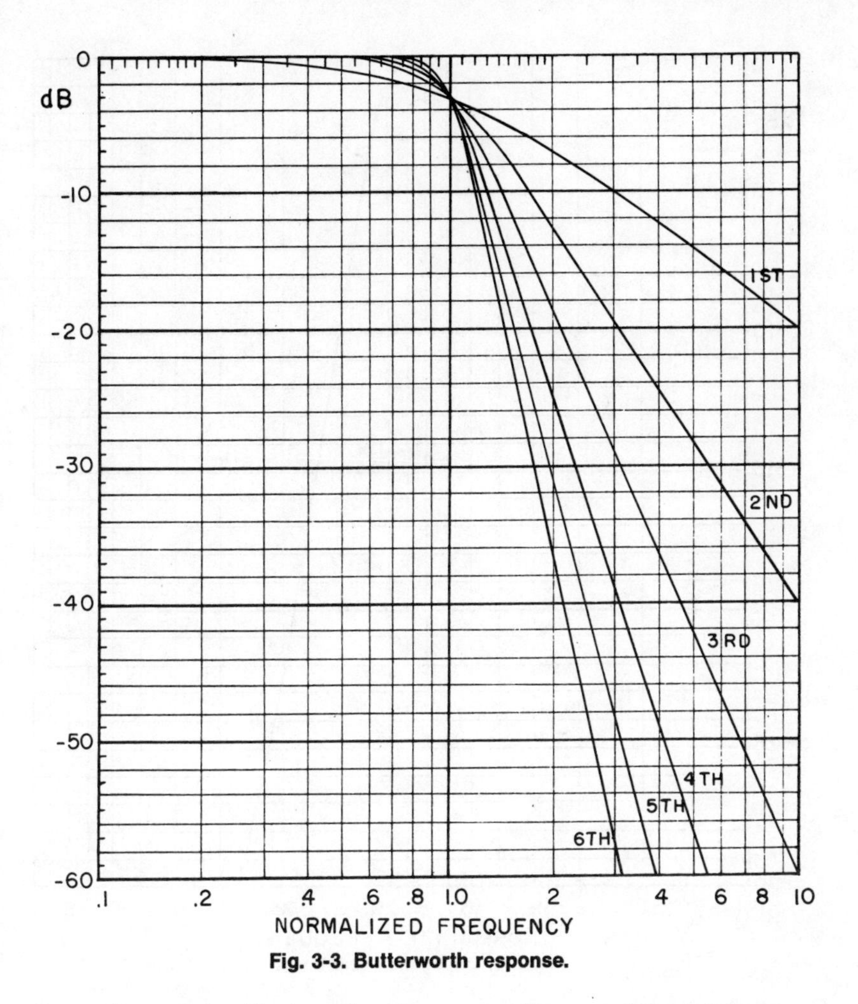

Fig. 3-3. Butterworth response.

Regardless of the filter's order, the curves shown in Fig. 3-3 have a constant, or *maximally flat passband,* which is called a *Butterworth response.* Alternatively, we may have an amplitude response shape that looks like the one shown in Fig. 3-4, having ripples in the passband. This type of response is characteristic of a *Chebyshev filter.* The one factor that determines the particular shape of the filter's passband shape is the *damping factor, or simply damping,* and is denoted by the symbol α. *The damping of a 2nd order filter section is a measure of its tendency to oscillate by itself.* As we will see in Chapters 5, 6, and 7, damping values will range from near zero to 2. The Butterworth response is an example of a *highly damped* filter, while the Chebyshev response illustrates a *slightly damped* filter. Although there are other filter shapes, we will present only the de-

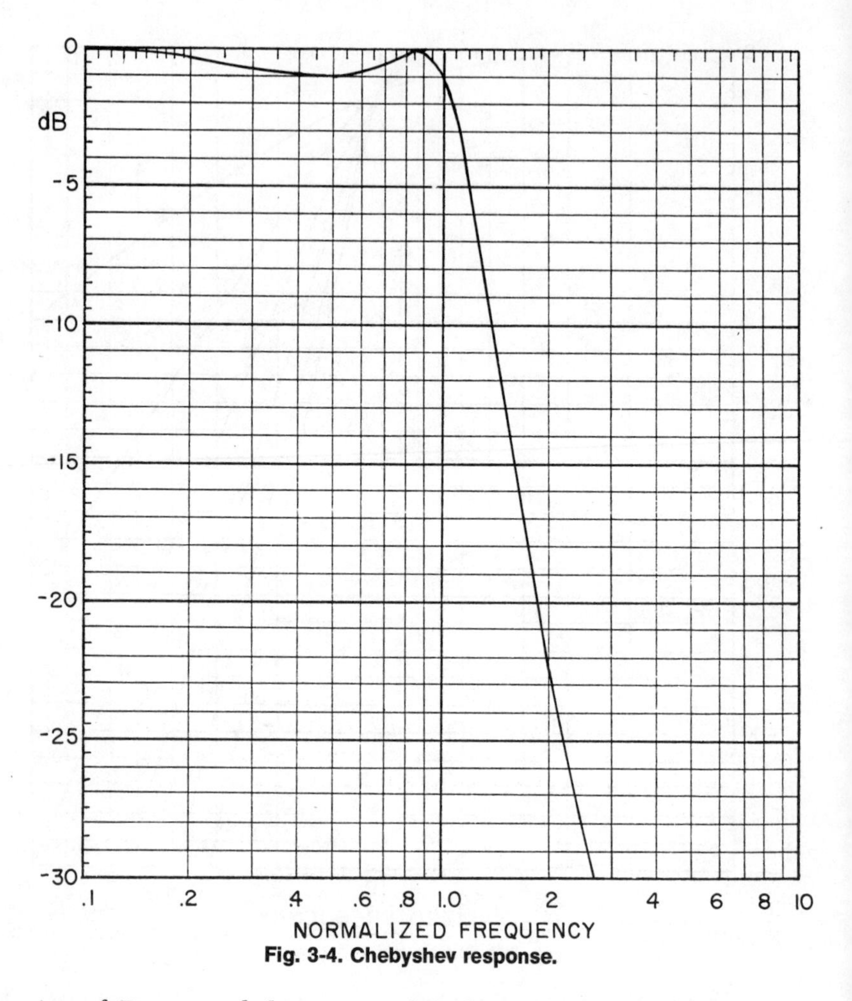

Fig. 3-4. Chebyshev response.

sign of Butterworth low-pass and high-pass filters. Other possible shape options are Bessel, Cauer, and Thompson.

THE HIGH-PASS FILTER

The generalized frequency response of a high-pass filter is simply the opposite of a low-pass filter, *rejecting signals below the cutoff frequency, and allowing signals above the cutoff frequency to pass through with little or no attenuation.*

As shown in Fig. 3-5, the amplitude response in the passband is essentially constant down to the cutoff frequency, where it rolls off at a rate that depends on the order of the filter. For a 2nd order high-pass filter, the rolloff is +12 dB/octave or +40 dB/decade. All of the

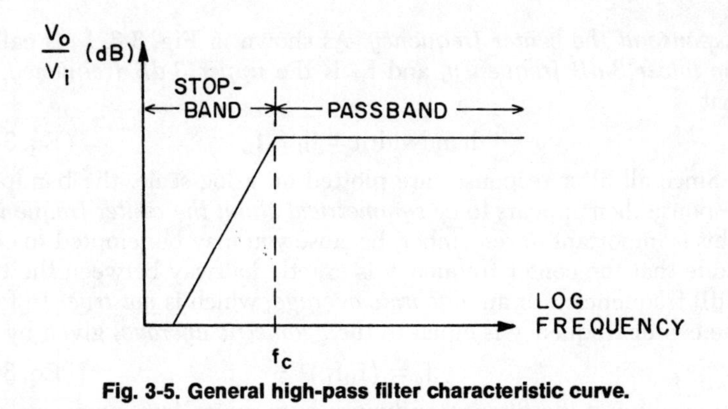

Fig. 3-5. General high-pass filter characteristic curve.

basic concepts that were applied to low-pass filters also apply to high-pass filters.

In reality, there can never be a true high-pass active filter, since the upper region of the filter's passband *is limited by the frequency response of the particular operational amplifier used,* as has been discussed in Chapter 2.

THE BANDPASS FILTER

As shown in Fig. 3-6, *a bandpass filter permits a range of median frequencies to pass through while rejecting frequencies above and below this median.* In defining the amplitude response of a bandpass filter, we are interested in its *center frequency and bandwidth.*

The center frequency (f_o) is in general, *the point where the maximum voltage gain occurs.* The bandwidth of a bandpass filter is *the difference between the upper and lower frequencies where the voltage gain is 0.707 times its maximum value, or 3 dB lower than the*

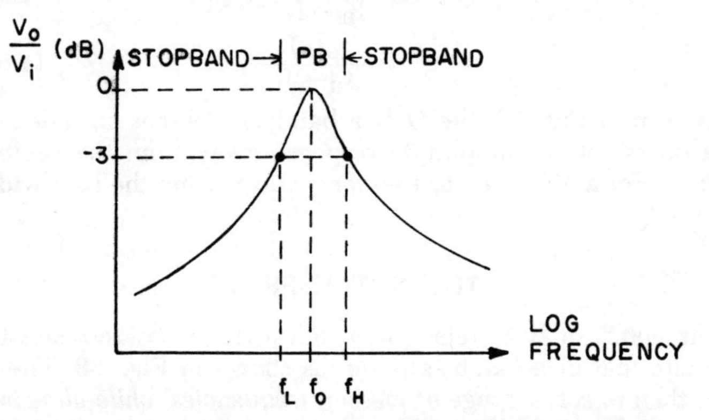

Fig. 3-6. Center frequency and bandpass of bandpass filters.

response at the center frequency. As shown in Fig. 3-6, f_L is called the *lower 3-dB frequency,* and f_H is the *upper 3-dB frequency,* so that

$$\text{bandwidth} = f_H - f_L \qquad \text{(Eq. 3-3)}$$

Since all filter responses are plotted on a log scale, the bandpass response then appears to be *symmetrical about the center frequency.* This is important to remember, because you may be tempted to conclude that the center frequency is exactly halfway between the two 3-dB frequencies, or an *algebraic average,* which is *not true!* In fact, the center frequency is equal to the *geometric average,* given by

$$f_o = (f_H f_L)^{1/2} \qquad \text{(Eq. 3-4)}$$

By specifying the filter's bandwidth and center frequency, we can then determine both 3-dB frequencies by

$$f_L = \frac{-BW + [(BW)^2 + (2f_o)^2]^{1/2}}{2} \qquad \text{(Eq. 3-5)}$$

and

$$f_H = F_L + BW \qquad \text{(Eq. 3-6)}$$

For example, suppose the center frequency is 1 kHz, and the bandwidth is 200 Hz. We may be strongly tempted to say the 3-dB frequencies are 900 and 1100 Hz. However, using Equation 3-4, we find that $f = 995$ Hz, which is not 1 kHz. However, by using Equations 3-5 and 3-6, $f_L = 905$ Hz, and $f_H = 1105$ Hz.

The bandpass filter's bandwidth and center frequency are related to each other by the *quality factor, or Q,* which is defined as

$$Q = \frac{f_o}{BW} \qquad \text{(Eq. 3-7)}$$

or

$$Q = \frac{f_o}{f_H - f_L} \qquad \text{(Eq. 3-8)}$$

$$= \frac{(f_H f_L)^{1/2}}{f_H - f_L} \qquad \text{(Eq. 3-9)}$$

As seen in Fig. 3-7, the Q of a bandpass filter is an index of the "sharpness" of the amplitude response away from the center frequency. For a given center frequency, decreasing the bandwidth increases Q.

THE NOTCH FILTER

The notch, or band-reject filter, has a frequency response that is opposite that of a bandpass filter, as shown in Fig. 3-8. The notch filter then *rejects a range of median frequencies while allowing frequencies above and below this range to pass through with little or no*

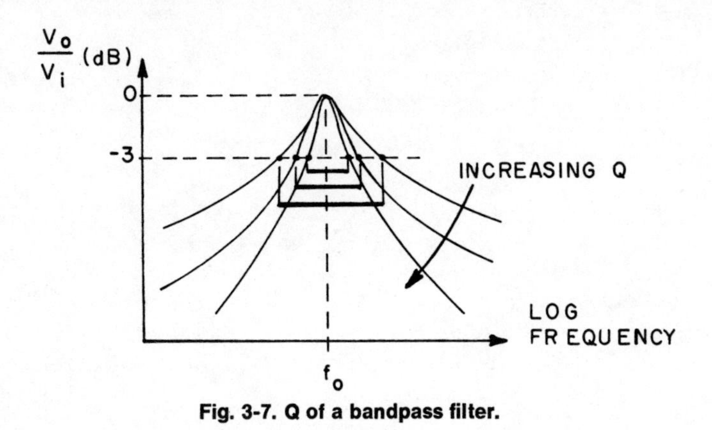

Fig. 3-7. Q of a bandpass filter.

attenuation. The definitions of bandwidth and Q are the same for notch filters as for bandpass filters.

The following chapters will individually discuss the design of the four filter classes presented in this chapter.

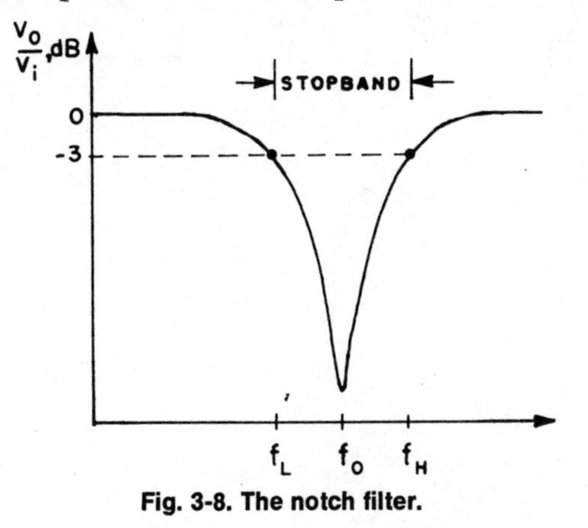

Fig. 3-8. The notch filter.

FREQUENCY

Throughout this book there will be two kinds of notations for frequency. The use of the letter "f" refers to the frequency expressed in hertz (Hz), such as in Equations 3-3, 3-4, etc. As will be used in the next chapter, the Greek letter ω (omega) is often used by electrical engineers as a sort of "mathematical trick" to make many calculations come out right. These two letters are related to each other by the factor of 2π, or 6.28, so that

$$\omega = 2\pi f \text{ (radians/s)} \qquad \text{(Eq. 3-10)}$$

Consequently, if any equation uses the variable "f," it can also be written in terms of ω by the use of the above equation.

CHAPTER 4

1st Order Low-Pass and High-Pass Active Filters

INTRODUCTION

In Chapters 2 and 3 we have been laying the necessary groundwork. In this chapter we start learning how to design 1st order low-pass and high-pass active filter sections.

OBJECTIVES

At the completion of this chapter, you will be able to do the following:

- Design and breadboard 1st order low-pass and high-pass filters
- Use simplified normalization and scaling techniques

1ST ORDER LOW-PASS ACTIVE FILTER

Fig. 4-1 shows the basic 1st order low-pass filter section. It is simply a combination of an RC low-pass *passive filter* with a *noninverting* amplifier. For $R = 1\Omega$, and $C = 1F$, the cutoff frequency is given by

$$\omega_c = 1/RC \qquad \text{(Eq. 4-1)}$$
$$= 1 \text{ radian/s}$$

For the frequency in terms of *hertz,* the relationship is

$$f_c = 1/6.28RC \qquad \text{(Eq. 4-2)}$$
$$= 0.159 \text{ Hz}$$

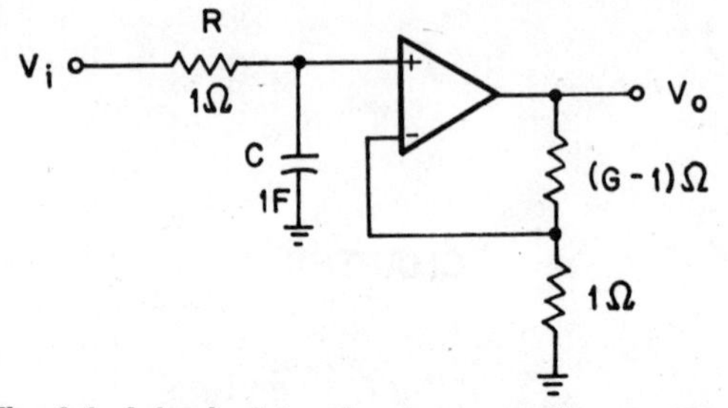

Fig. 4-1. A basic 1st order, low-pass filter section.

Since the circuit of Fig. 4-1 has a cutoff frequency of 1 radian/s (0.16 Hz), we say that this circuit is *normalized at 1 radian/s*, so that this is our starting point. The passband gain is determined solely by the feedback resistance, which is equal to (G − 1), when the op-amp input resistance is 1Ω.

Without subjecting you to extensive derivations, the amplitude vs. frequency response of this 1st order low-pass filter is given by

$$dB = 20 \log_{10} G - 20 \log_{10} (1 + \omega^2)^{1/2} \qquad \text{(Eq. 4-3)}$$

where ω is the frequency of interest *divided by the cutoff frequency*. If the passband gain is unity (0 dB), then Equation 4-3 becomes

$$dB = - 20 \log_{10} (1 + \omega^2)^{1/2} \qquad \text{(Eq. 4-4)}$$

Fig. 4-2 shows the plot of the voltage gain of a 1st order low-pass filter, given by Equation 4-3. It should be noted that this graph is

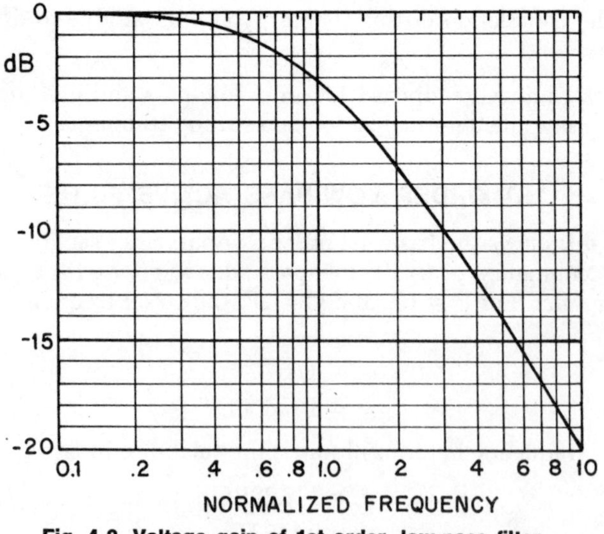

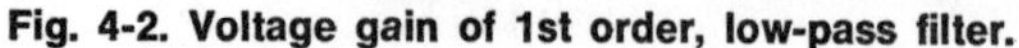

Fig. 4-2. Voltage gain of 1st order, low-pass filter.

normalized for a cutoff frequency of 1.0 and a passband gain of 0 dB. In addition, the response is from one decade below ($\omega = 0.1$) to one decade above ($\omega = 10$) the cutoff frequency. In the next section, we shall describe the usefulness of such a graph.

SCALING

In order to make the graph of Fig. 4-2 useful for any cutoff frequency, we introduce the concept of *scaling.** The *analysis* of any filter section is done on a circuit having a cutoff frequency of 1 radian/s and an impedance level of 1Ω, such as the low-pass section of Fig. 4-1. Alternatively, this circuit can be based on a *1-Hz cutoff frequency,* using Fig. 4-3.

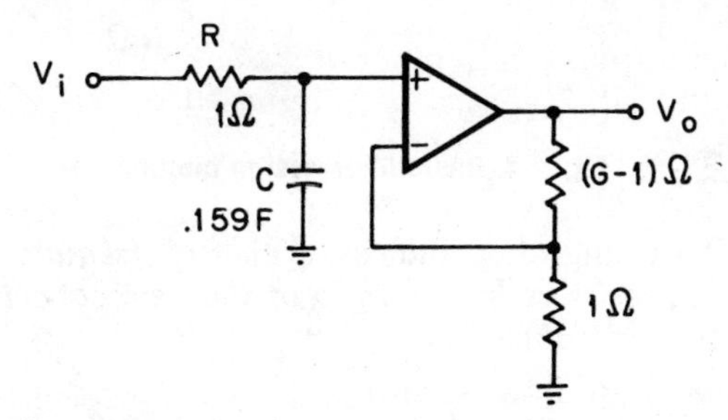

Fig. 4-3. Low-pass section with cutoff frequency of 1 radian/second.

Rule #1: To change the cutoff frequency of a circuit, either *multiply* all frequency-determining resistors by the ratio of the *old* frequency to the *new* frequency, or multiply all frequency-determining capacitors by this ratio; *never do both!*

The *design* of any filter section is done on a circuit having a 10-kΩ impedance level and a 1-kHz cutoff frequency, using the circuit shown in Fig. 4-4.

Example:

Convert the 1-kHz circuit of Fig. 4-4 to a cutoff frequency of 2.5 kHz.

First, either multiply the resistor by 0.4 (i.e., 1 kHz/2.5 kHz), *or* multiply the 0.0159-μF capacitor by 0.4, as shown in Fig. 4-5 so that

---

* One good book outlining this method is: Lancaster, D. *Active-Filter Cookbook.* Howard W. Sams & Co., Inc., Indianapolis, 1975.

$$f_c = \frac{1}{(6.28)(.0159\ \mu F)(4\ k\Omega)}$$
$$= 2.5\ kHz$$

As seen from Fig. 4-5, .0159 μF is rather hard to find, since it is not a *standard capacitor value* (see Appendix A). To change this capacitor to a standard value, we introduce the concept of *impedance scaling*.

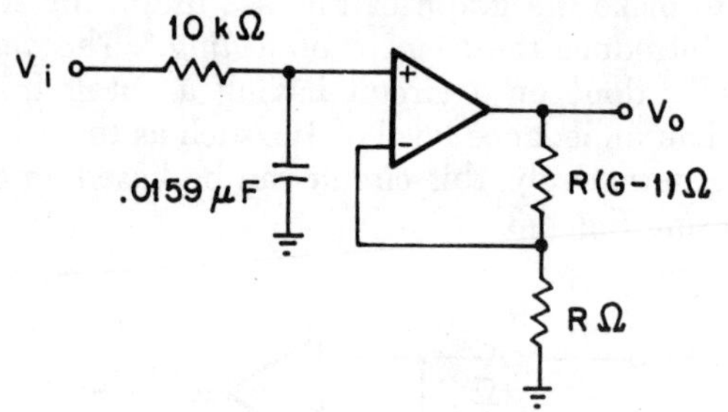

Fig. 4-4. Basic filter design circuit.

Rule #2: To change impedances, *divide all frequency-determining resistors by the ratio of the new capacitor value to the old value.*

To change the .0159-μF capacitor to a .022-μF capacitor, for example, we divide the 4 kΩ resistor by 1.38 (i.e., .022/.0159), giving a 2.9-kΩ resistor, and the circuit of Fig. 4-6. Using Equation 4-2 we can check ourselves,

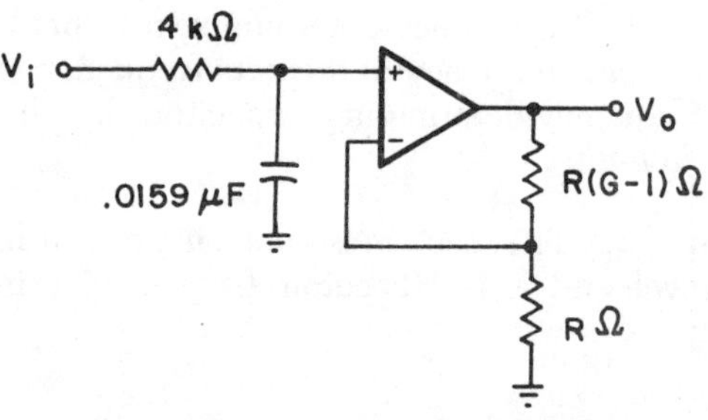

Fig. 4-5. Converted circuit of Fig. 4-4.

$$f_c = \frac{1}{(6.28)(2.9\ k\Omega)(.022\ \mu F)}$$
$$= 2.5\ kHz$$

Remember: It should be emphasized that the circuits of Figs. 4-5 and 4-6 are equivalent; that is, *both circuits have a cutoff frequency of 2.5 kHz.*

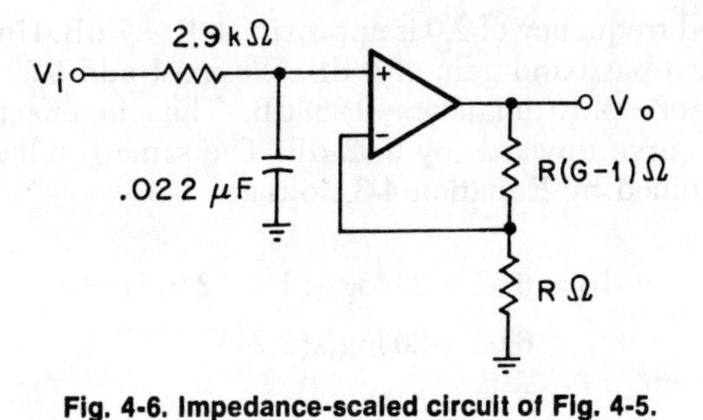

Fig. 4-6. Impedance-scaled circuit of Fig. 4-5.

The voltage gain, or amplitude response vs. frequency for this 2.5-kHz low-pass filter can be easily found from the normalized graph of Fig. 4-2. Suppose that the passband gain of this filter section is 6.02 dB (equal to a voltage ratio of 2.0). Taking Fig. 4-2, we multiply the horizontal axis by 2.5 kHz so that the response will cover the range from 250 Hz (one decade below the 2.5-kHz cutoff) to 25 kHz (one decade above). In addition we add 6.02 dB to the vertical axis, as shown in Fig. 4-7.

However, instead of redrawing the response curve each time we select a different cutoff frequency, we can just mentally change Fig. 4-2. For example, what will be the amplitude response at 5.0 kHz? If we first normalize the 2.5-kHz cutoff frequency to 1.0, then a frequency of 5.0 kHz corresponds to a normalized frequency of 2.0 (i.e., 5.0/2.5). From Fig. 4-2, we see that the amplitude response at

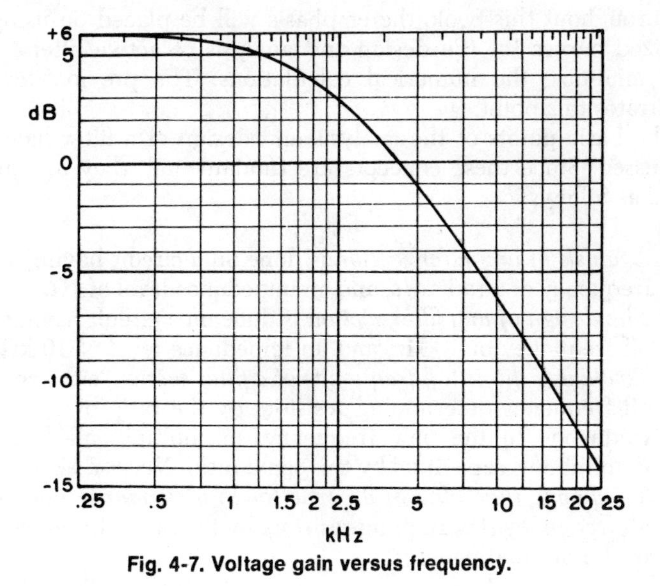

Fig. 4-7. Voltage gain versus frequency.

a normalized frequency of 2.0 is approximately −7 dB. However, this graph is for a passband gain of 0 dB. We must add 6.02 dB, giving a response of approximately −0.98 dB. Thus in effect, we have *shifted* the curve upwards by 6.02 dB. The same result could have been determined by Equation 4-3, so that

$$\begin{aligned} dB &= 6.02 - 20 \log_{10}(1 + (2.0)^2)^{1/2} \\ &= 6.02 - 20 \log_{10}(2.24) \\ &= 6.02 - 6.99 \\ &= -0.97 \text{ dB} \end{aligned}$$

Once you have practiced using only the normalized response curves, the complicated equations can then be ignored.

To make it easier to determine the necessary standard resistor and capacitor values for any cutoff frequency, we can use the graph of Fig. 4-8. instead of performing impedance scaling calculations. For the 2.5-kHz low-pass filter, first locate 2.5 kHz on the horizontal "frequency" axis. Then go straight up until you intersect one of the horizontal lines representing the standard capacitor values. If for example, using a standard value of .022 μF, we see that the resistance required for a 2.5-kHz cutoff frequency lies between 2.7 kΩ and 3.3 kΩ. However, since this point is closer to the 2.7 kΩ line, we use this value. We also could have chosen a .0022-μF capacitor and 27-kΩ resistor.

Throughout this book, the emphasis will be placed on using normalized curves for the design and analysis of active filters, which will minimize the numerical calculations. The previous example illustrates this point.

The basic points of the analysis and design of a filter have been discussed. Since these concepts are fundamental, they are summarized as follows:

- *Analysis of any filter section* is done on a circuit having a cutoff frequency of 1 radian/s, and an impedance level of 1Ω.
- *The design of any filter section* is done on a circuit having a cutoff frequency of 1 kHz, and an impedance level of 10 kΩ.
- *To change the cutoff frequency of a filter section,* either multiply all frequency-determining resistors by the ratio of the original frequency to the new frequency, or multiply all frequency-determining capacitors by this same ratio. *Never do both!*
- *To change the value of a capacitor to a standard value,* divide all frequency-determining resistors by the ratio of the new value to the original value

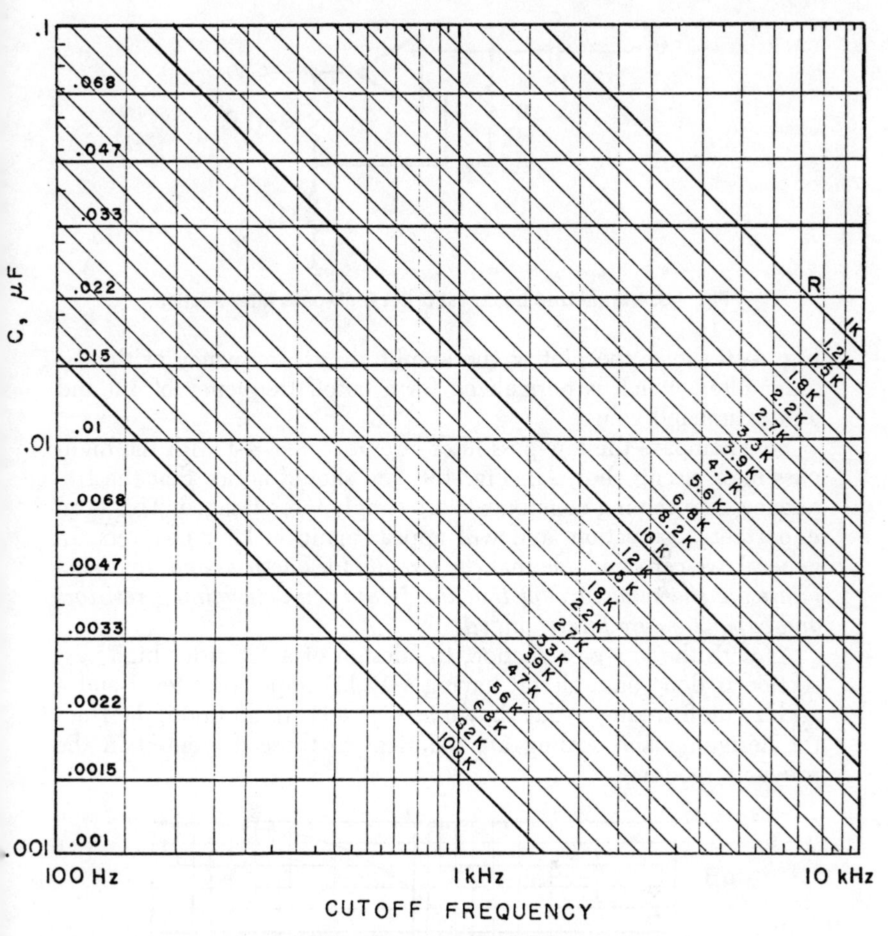

Fig. 4-8. RC values versus cutoff frequency.

1ST ORDER HIGH-PASS ACTIVE FILTER

Fig. 4-9 shows the basic 1st order high-pass filter section, which is normalized for a cutoff frequency of 1 radian/s, given by Equation 4-1. For a passband gain G, the amplitude vs. frequency response of this 1st order high-pass section is given by

$$\mathrm{dB} = 20 \log_{10}\mathrm{G} + 20 \log_{10}\left[\frac{\omega}{(1+\omega^2)^{1/2}}\right] \qquad \text{(Eq. 4-5)}$$

If the passband gain is unity (0 dB), then Equation 4-5 becomes

$$\mathrm{dB} = 20 \log_{10}\left[\frac{\omega}{(1+\omega^2)^{1/2}}\right] \qquad \text{(Eq. 4-6)}$$

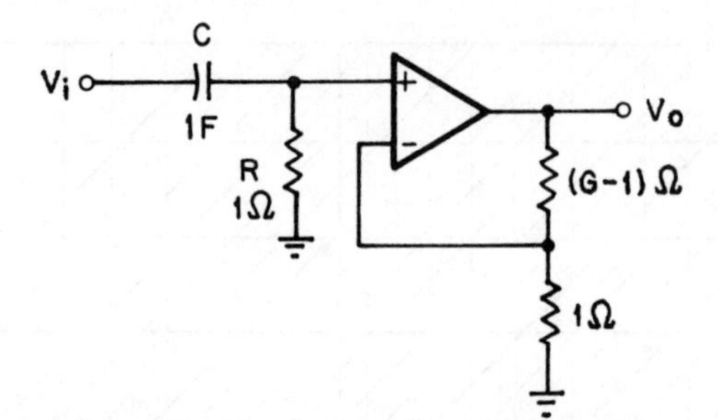

Fig. 4-9. Normalized basic 1st order high-pass filter section.

Fig. 4-10 shows the plot of the amplitude vs. frequency of the 1st order filter, which is normalized for a cutoff frequency of 1.0 and unity passband gain.

Now compare the low-pass filter circuit of Fig. 4-1 with the high-pass filter circuit shown in Fig. 4-9. You should notice that the frequency-determining components are simply interchanged. This is an important observation, and will appear again in later chapters. In general, we can conclude that *the circuits for low-pass and high-pass filters are the same, except that the frequency-determining resistors and capacitors are interchanged.*

As with the low-pass section, the design of a 1st order high-pass section is done on a circuit having a 10 kΩ impedance level and a 1-kHz cutoff frequency, as shown in Fig. 4-11. In addition, the rules for designing and scaling are identical to those presented in the previous section.

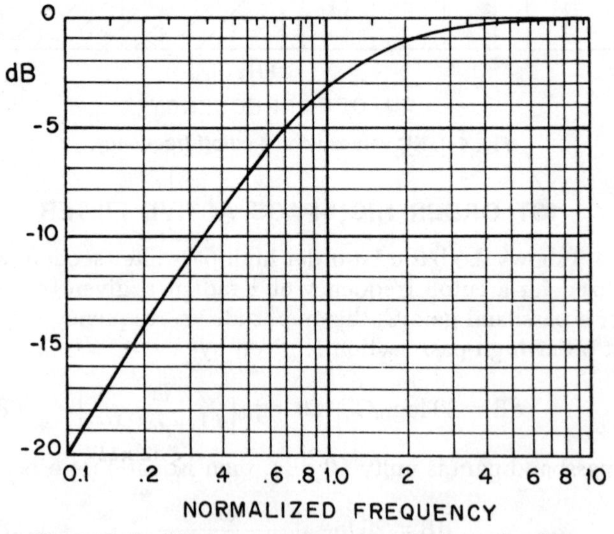

Fig. 4-10. Amplitude versus frequency for 1st order filter.

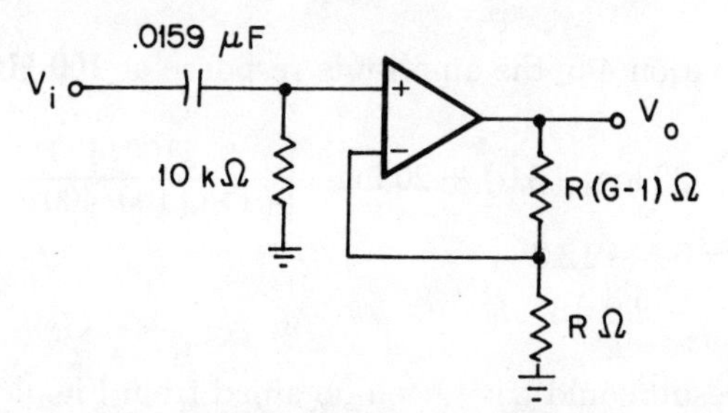

Fig. 4-11. Basic design of 1st order high-pass section.

For both the low-pass and high-pass filter sections, we have used a *noninverting* amplifier following the resistor-capacitor network. Consequently the passband voltage gain will always be greater than 1 (or 0 dB). If unity passband gain is required, a voltage follower is used after the resistor-capacitor network, as illustrated in Fig. 4-12 for a 1st order high-pass filter.

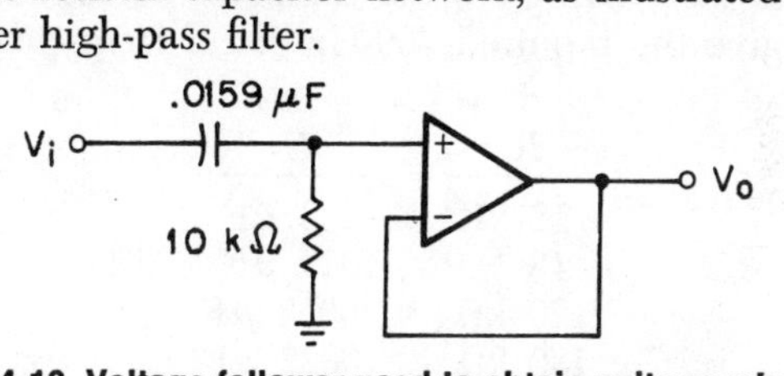

Fig. 4-12. Voltage follower used to obtain unity passband.

Example:

Design a 400-Hz 1st order high-pass filter with a passband voltage gain of 3.0 (9.5 dB), and determine the amplitude response at 100 Hz.

Using the circuit of Fig. 4-11, first scale the cutoff frequency to 400 Hz by multiplying the 10 kΩ by 2.5 (i.e., 1000/400). Next, the capacitor should be converted to a standard value, for example .033 μF. Therefore we divide the resistor by 2.075 (i.e., .033/.0159), resulting in the circuit given in Fig. 4-13.

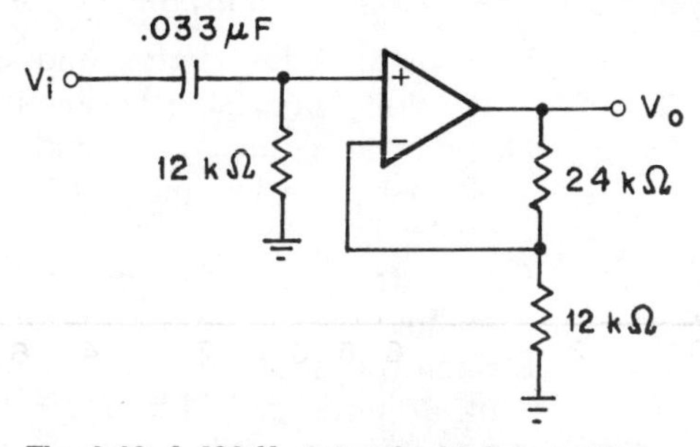

Fig. 4-13. A 400-Hz 1st order high-pass filter.

Using Equation 4-5, the amplitude response at 100 Hz will be

$$\text{dB} = 20 \log_{10}(3.0) + 20 \log_{10}\left[\frac{(100/400)}{(1 + [100/400]^2)^{1/2}}\right]$$

$$= 9.5 - 12.3$$

$$= -2.8 \text{ dB}$$

The above result could have been obtained from Fig. 4-10. The amplitude response at a normalized frequency of 0.25 (i.e., 100/400) is approximately −12 dB. Since this graph has been normalized for a passband gain of 0 dB, we must *add* 9.5 dB to −12 dB to give an overall response of approximately −2.5 dB.

Instead of calculating the required standard resistor-capacitor combination to give a cutoff frequency of 400 Hz, we can easily refer to Fig. 4-8. The following resistor-capacitor combinations will approximately give the required 400-Hz cutoff frequency:

R	*C*
12 kΩ	.033 μF
18 kΩ	.022 μF
120 kΩ	.0033 μF
180 kΩ	.0022 μF

AN INTRODUCTION TO THE EXPERIMENTS

The following experiments are designed to demonstrate the design and operation of 1st order active filters. In addition, the basic experimental technique applicable for determining the amplitude response of any filter, will be demonstrated.

The experiments that you will perform can be summarized as follows:

Experiment No.	*Purpose*
1	Demonstrates the design and operation of a unity gain 1st order low-pass filter.
2	Demonstrates the design and operation of a 1st order low-pass filter with positive gain.
3	Demonstrates the concept of frequency scaling.
4	Demonstrates the design and operation of a unity gain 1st order high-pass filter.

EXPERIMENT NO. 1

Purpose

The purpose of this experiment is to demonstrate the operation of a unity gain 1st order low-pass active filter.

Schematic Diagram of Circuit (Fig. 4-14)

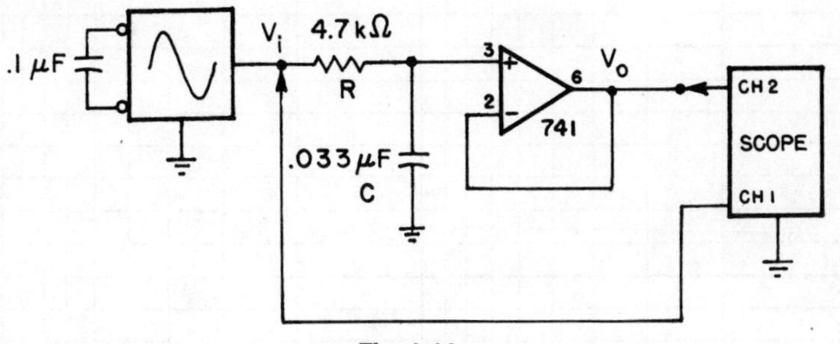

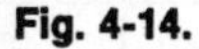

Fig. 4-14.

Design Basics

- Cutoff Frequency: $f_c = 1/2\pi RC$
- Amplitude Response: $20 \log_{10} \left[\frac{1}{(1 + f^2)}\right]^{1/2}$

Step 1

Set your oscilloscope for the following settings:

- Channels 1 & 2: 200 mV/division
- Time base: 1 ms/division
- AC coupling

Step 2

Wire the circuit shown in the schematic diagram (don't forget the power supply connections to the op amp!). Apply power to the breadboard and adjust the input signal to 1 volt peak-to-peak (5 full divisions). Also adjust the generator's input frequency to 100 Hz (1 full cycle/10 divisions).

Step 3

From the equation given in the "design basics" section of this experiment, what do you expect the cutoff frequency to be?

Your answer should be approximately 1026 Hz. Practically, we'll assume the cutoff frequency to be 1000 Hz.

Step 4

By means of the generator's FREQUENCY control, vary the frequency to complete the following table, and plot your results on the blank graph (Fig. 4-15) provided for this purpose.

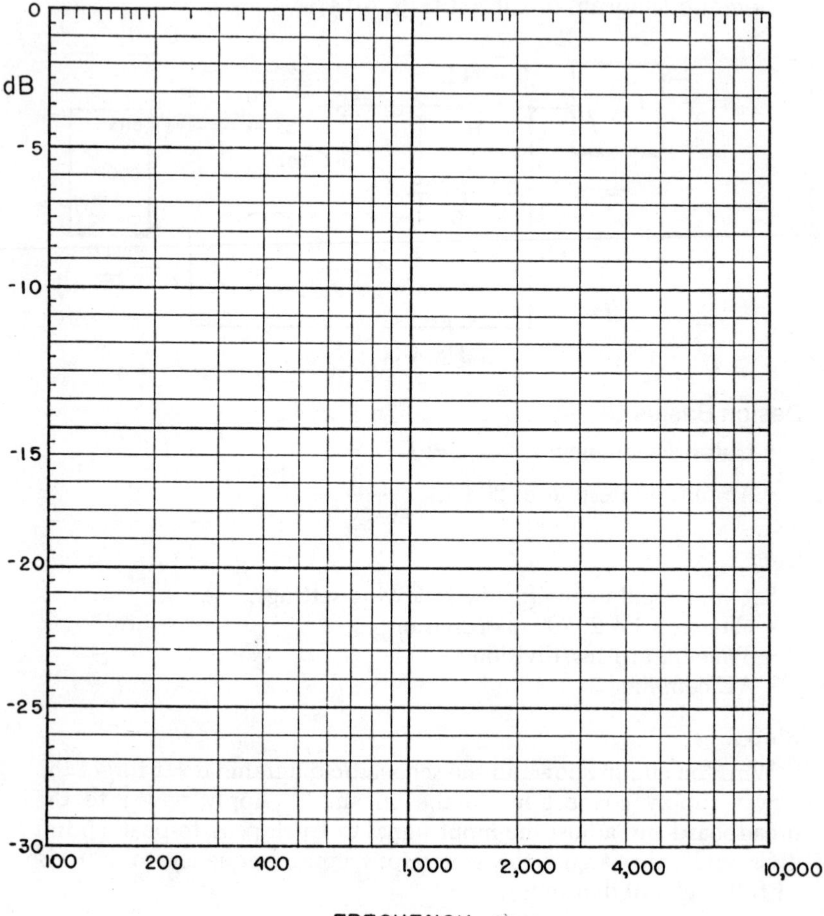

Fig. 4-15.

Note: Because of the simplicity of the LR-31 Outboard, the output voltage will decrease slightly with increasing frequency. Be careful to check that the input voltage is 1.0 volt for *each* measurement. Remember, the dB gain is calculated from the equation:

$$dB = 20 \log_{10} \frac{V_o}{V_i}$$

at each measured frequency.

In addition, the expected theoretical gain for each frequency is listed to compare your experimental findings.

Frequency	V_o	V_o/V_i	Experimental dB Gain	Theoretical dB Gain
100 Hz				−0.04
200				−0.17
300				−0.37
400				−0.64
500				−0.97
600				−1.34
700				−1.73
800				−2.15
900				−2.58
1000				−3.01
2000				−6.99
4000				−12.3
10,000*				−20.0

* Using the LR-31 Outboard, it may be necessary to use a capacitor value that is less than .1 μF to obtain a 10-kHz signal. If the output is zero or distorted, use a .01-μF capacitor to increase the frequency range to approximately 1 kHz to 100 kHz.

As a comparison, my experimental results are presented in the following table, and these results are plotted along with the theoretical response in Fig. 4-16. The difference between the two curves is probably attributed to component tolerance, unless a measurement error has been made, such as forgetting to check that the input voltage is constant over the 100 Hz to 10 kHz measurement range. I used a 5% resistor and a 10% capacitor, and, at the worst, the cutoff frequency will be 880 Hz or 1200 Hz, or somewhere in between, as shown on the graph.

Frequency	V_o	V_o/V_i	Experimental dB Gain	Theoretical dB Gain
100 Hz	1.00 V	1.00	0	−0.04
200	1.00	1.00	0	−0.17
300	0.98	0.98	−0.18	−0.37
400	0.94	0.94	−0.54	−0.64
500	0.90	0.90	−0.92	−0.97
600	0.86	0.86	−1.31	−1.34
700	0.80	0.80	−1.94	−1.73
800	0.76	0.76	−2.38	−2.15
900	0.72	0.72	−2.85	−2.58
1000	0.68	0.68	−3.35	−3.01
2000	0.40	0.40	−7.96	−6.99
4000	0.22	0.22	−13.2	−12.3
10,000	0.08	0.08	−21.9	−20.0

EXPERIMENT NO. 2

Purpose

The purpose of this experiment is to demonstrate the operation of a 1st order low-pass active filter having positive gain.

Schematic Diagram of Circuit (Fig. 4-17)

Design Basics

- Cutoff Frequency: $f_c = 1/2\pi RC$
- Gain: $G = 1 + R_B/R_A$
- Amplitude Response: $20 \log_{10}\left[\frac{G}{(1+f^2)^{1/2}}\right]$

Step 1

Set your oscilloscope for the following settings:

- Channel 1: 200 mV/division
- Channel 2: 500 mV/division
- Time base: 1 ms/division
- AC coupling

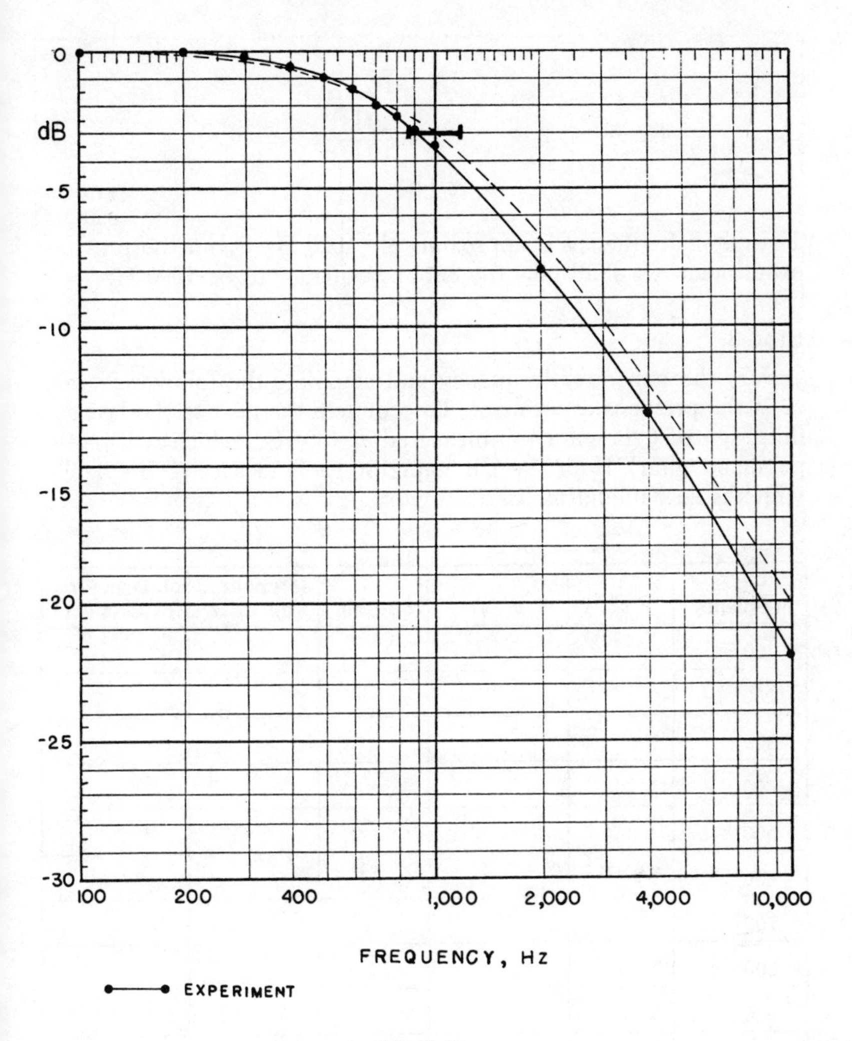

Fig. 4-16.

Step 2

Wire the circuit in the schematic diagram. Apply power to the breadboard and adjust the input voltage to 1 volt peak-to-peak, and the frequency to 100 Hz. With the circuit values shown, what should be the passband gain and the cutoff frequency be?

The passband gain is 2.0 (+6.02 dB), since

$$G = 1 + \frac{R_B}{R_A}$$

$$= 1 + \frac{10\ k\Omega}{10\ k\Omega}$$

$$= 2.0$$

The cutoff frequency is approximately 1026 Hz. As in the previous experiment, we shall take the cutoff frequency to be 1000 Hz.

Step 3

Vary the generator frequency and complete the following table, as in the previous experiment. To complete the next to the last column, subtract the gain you measured at 100 Hz (which will be the passband gain) from the dB gain for each measured frequency, which is now the *normalized dB gain.*

Frequency	V_o	V_o/V_i	dB Gain	Normalized dB Gain	dB Gain From Experiment I
100 Hz					
200					
300					
400					
500					
600					
700					
800					
900					
1000					
2000					
4000					
10,000					

Step 4

On the blank graph page (Fig. 4-18), plot the results from the last two columns in Step 3. Are the two curves similar?

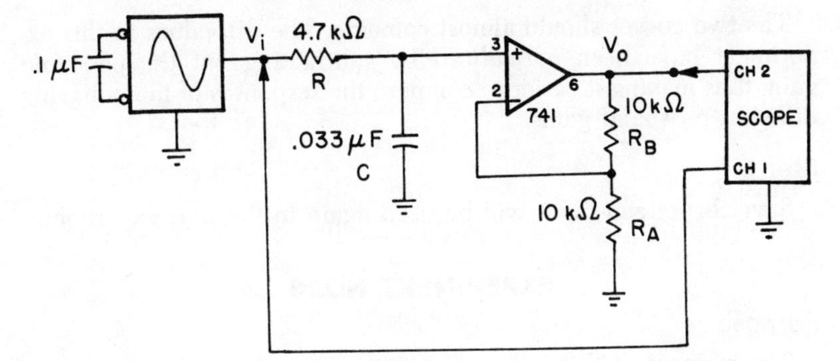

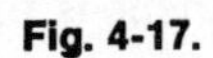

Fig. 4-17.

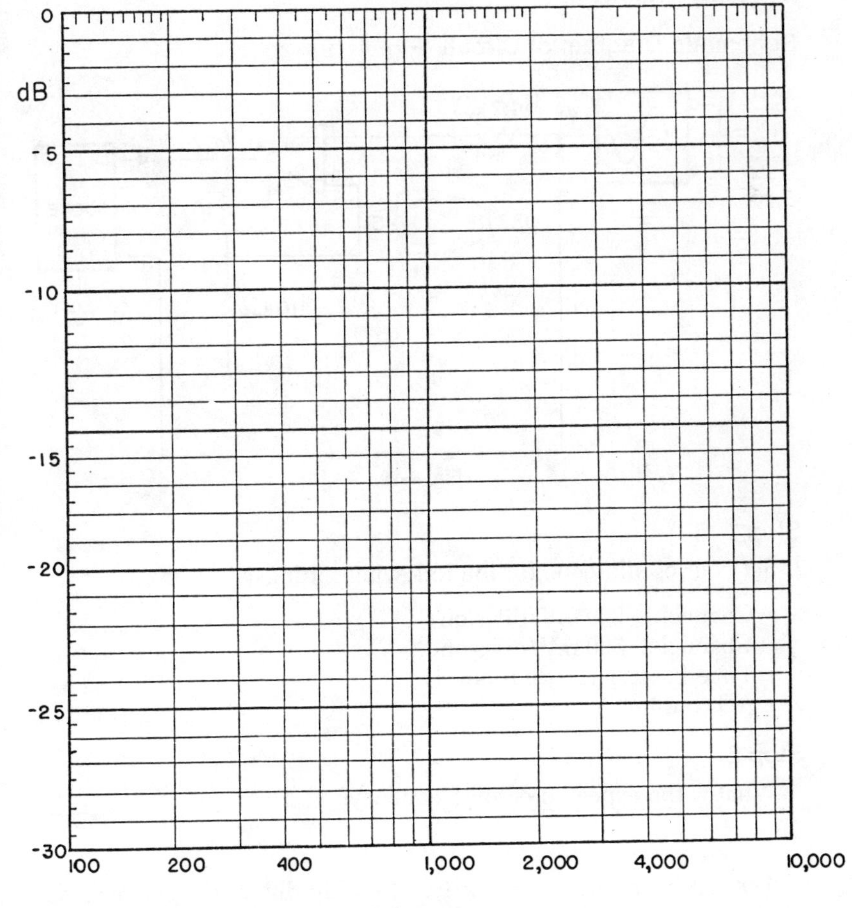

Fig. 4-18.

The two curves should almost coincide. The dB values in this experiment have been *normalized* by subtracting out the passband gain, thus making it easier to compare the responses of filters having different passband gains.

Step 5

Save this circuit, as it will be used again in the next experiment.

EXPERIMENT NO. 3

Purpose

The purpose of this experiment is to demonstrate the concept of frequency scaling.

Schematic Diagram of Circuit (Fig. 4-19)

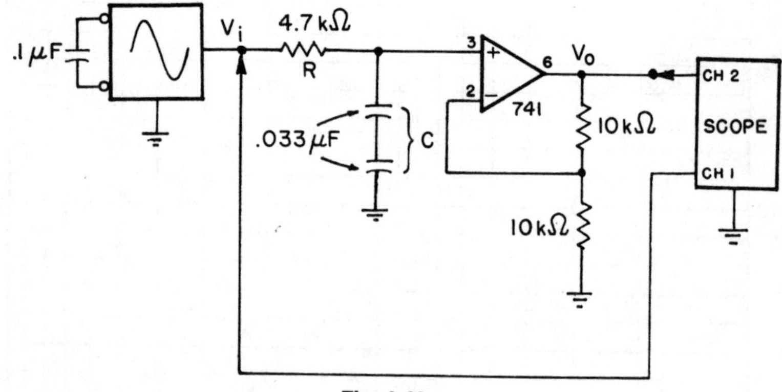

Fig. 4-19.

Step 1

Set your oscilloscope for the following settings:

- Channel 1: 200 mV/division
- Channel 2: 500 mV/division
- Time base: 1 ms/division
- AC coupling

Step 2

What is the expected cutoff frequency?

The cutoff frequency is 2 kHz. The only difference between the circuit of this experiment and that of Experiment No. 2, is that there are *two 0.033-μF capacitors in series*. Consequently, the frequency-determining capacitance is now *halved,* or 0.0165 μF, and the cutoff frequency will be *twice the previous value, or 2 kHz. It is much easier to connect these two capacitors in series than trying to find a 0.0165-μF capacitor!*

Step 3

Wire the circuit in the schematic diagram. Apply power to the breadboard and adjust the input signal to 1 volt peak-to-peak and a frequency of 200 Hz, completing the following table and plotting your results on the blank graph provided for this purpose in Fig. 4-20.

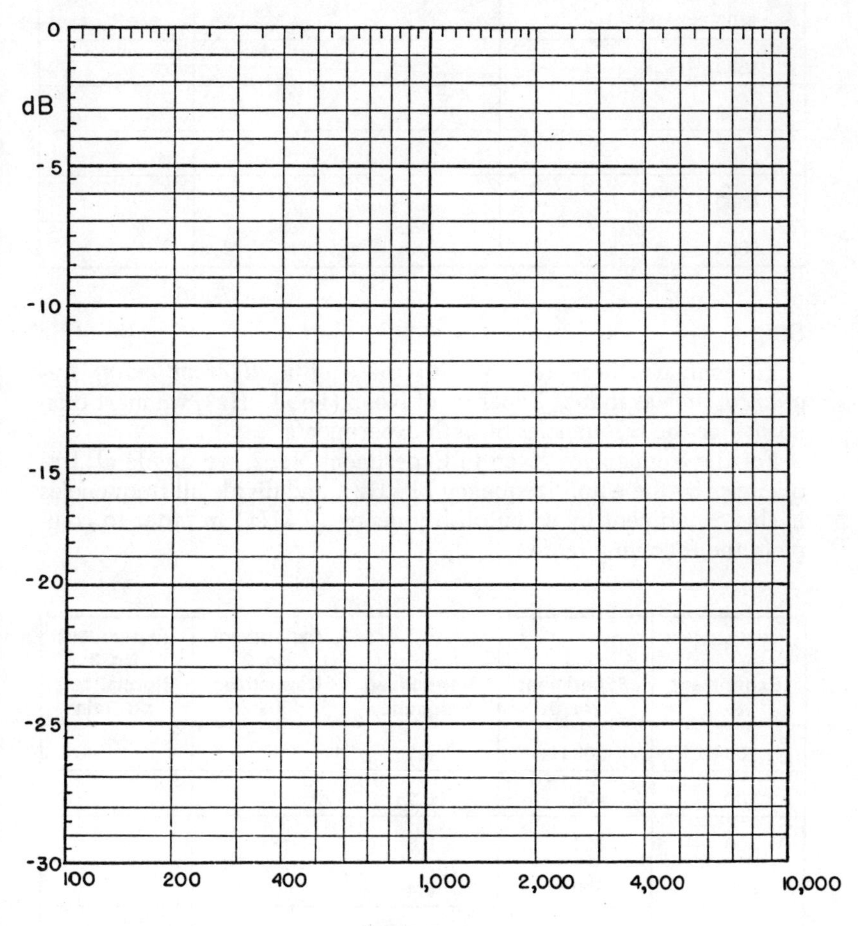

Fig. 4-20.

Frequency	V_o	V_o/V_i	dB Gain	Normalized dB Gain
200 Hz				
400				
600				
800				
1000				
1200				
1400				
1600				
1800				
2000				
4000				
8000				

Step 4

To compare these results with those of a *different cutoff frequency,* such as that of Experiment No. 2 (i.e., 1 kHz), we must then *normalize* the frequencies of both experiments.

For the frequencies given in Experiment No. 2, we divide all frequencies by the cutoff frequency (1 kHz), and divide all frequencies in this experiment by its cutoff frequency (2 kHz) in order to complete the following table.

Experiment No. 2	Experiment No. 3	Normalized Frequency	Experiment No. 2 Normalized dB Gain	Experiment No. 3 Normalized dB Gain
100 Hz	200 Hz	0.1		
200	400	0.2		
300	600	0.3		
400	800	0.4		
500	1000	0.5		
600	1200	0.6		

Experiment No. 2	Experiment No. 3	Normalized Frequency	Experiment No. 2 Normalized dB Gain	Experiment No. 3 Normalized dB Gain
700	1400	0.7		
800	1600	0.8		
900	1800	0.9		
1000	2000	1.0 (cutoff)		
2000	4000	2.0		
4000	8000	4.0		

If you have completed the above table correctly, you should see that the results in the last two columns are approximately the same. Therefore, we have now just compared the responses of two low-pass filters with different cutoff frequencies.

Step 5

As an optional exercise, remove one of the .033-μF capacitors from the circuit, grounding the remaining lead, and place a 4.7-kΩ resistor in parallel with the 4.7 kΩ resistor already in the circuit, as shown in the partial diagram of Fig. 4-21. Now repeat Steps 3 and 4 of this experiment. Do all three results agree?

Since all we did was to double the cutoff frequency of Experiment No. 2 by halving the frequency-determining resistance, instead of halving the frequency determining capacitance, the results should be approximately the same.

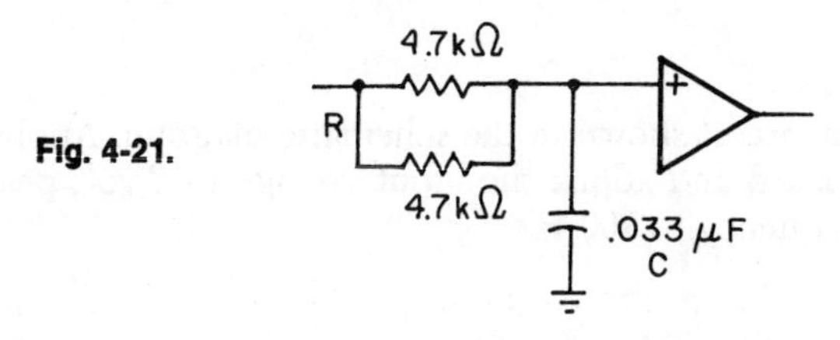

Fig. 4-21.

If the resistor is halved, and the capacitor is doubled, what is the cutoff frequency?

There will be no change!

EXPERIMENT NO. 4

Purpose

The purpose of this experiment is to demonstrate the operation of a unity gain 1st order high-pass active filter.

Schematic Diagram of Circuit (Fig. 4-22)

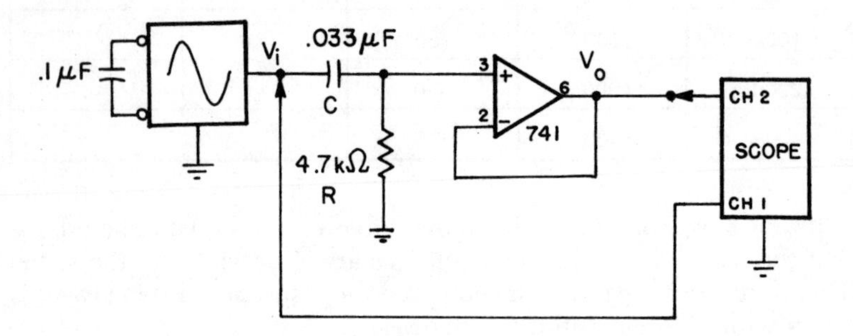

Fig. 4-22.

Design Basics

- Cutoff Frequency: $f_c = 1/2\pi RC$
- Amplitude Response: $20 \log_{10}\left[\frac{f}{(1+f^2)^{1/2}}\right]$

Step 1

Set your oscilloscope for the following settings:

- Channels 1 & 2: 200 mV/division
- Time base: 1 ms/division
- AC coupling

Step 2

Wire the circuit shown in the schematic diagram. Apply power to the breadboard and adjust the input voltage to 1 volt peak-to-peak, and the frequency to 100 Hz.

Step 3

Vary the generator frequency to complete the following table, plotting your results on the blank graph provided for this purpose in Fig. 4-23.

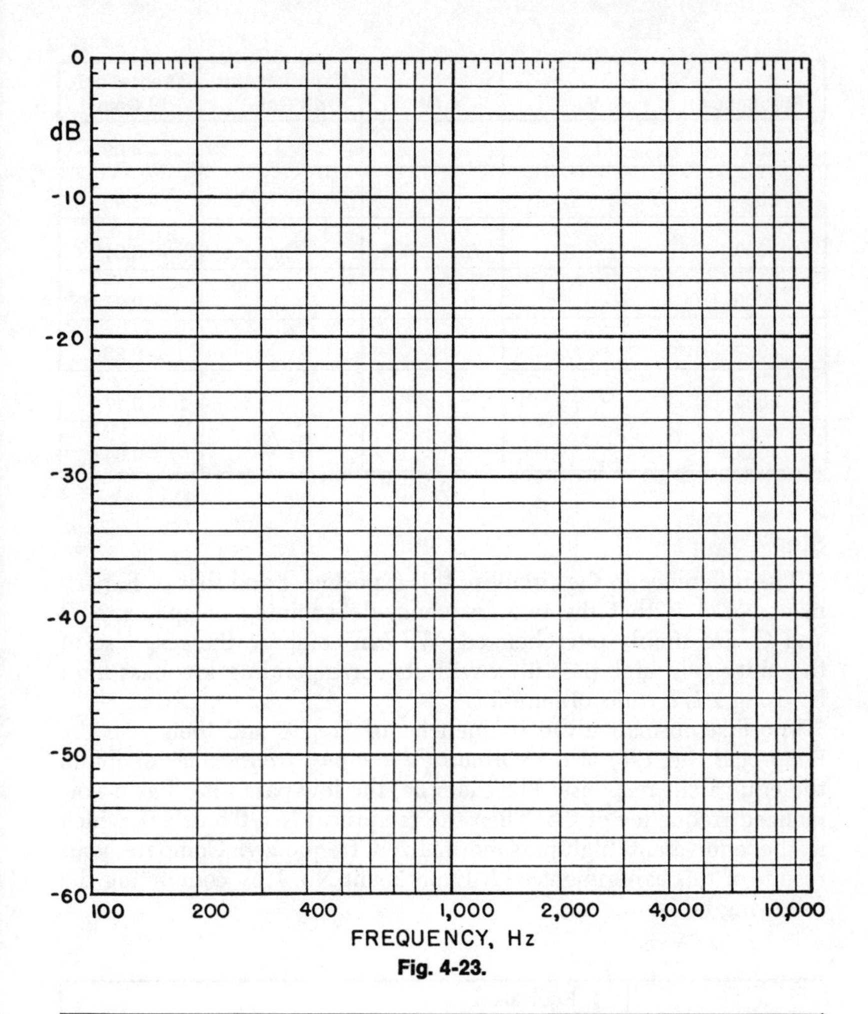

Fig. 4-23.

Frequency	V_o	V_o/V_1	Experimental dB Gain	Theoretical dB Gain
100				−20.2
200				−14.2
300				−10.8
400				−8.60
500				−6.99
600				−5.77
700				−4.83

Frequency	V_o	V_o/V_i	Experimental dB Gain	Theoretical dB Gain
800				—4.09
900				—3.49
1000				—3.01
2000				—0.97
4000				—0.26
5000				—0.17
10,000				—0.04

Step 4

The difference in the circuit of this experiment and that of Experiment No. 1 is that the two frequency-determining components, R and C, are simply interchanged. We can compare the response of the unity gain high-pass filter with its corresponding low-pass filter by using a *1/F transformation.*

We first normalize the frequencies of Step 3 and then take the *reciprocal* (i.e, 1/f) of the *normalized low-pass frequencies* to obtain the equivalent response. For example, the low-pass filter has a normalized frequency of 0.5. Then the reciprocal is 1/0.5 or 2.0, which is the equivalent high-pass normalized frequency. Compare your results of this experiment with Experiment No. 1 by completing the following table.

Normalized Frequency Low-Pass	Equivalent Normalized Frequency High-Pass (1/f)	Experiment No. 1 dB Gain	Experiment No. 4 dB Gain
0.1 (100 Hz)	1.0 (1000 Hz)		
0.2 (200 Hz)	5.0 (5000 Hz)		
0.5 (500 Hz)	2.0 (2000 Hz)		
1.0 (1000 Hz)	10.0 (1000 Hz)		
2.0 (2000 Hz)	0.5 (500 Hz)		
10.0 (10,000 Hz)	0.1 (100 Hz)		

Consequently, by knowing the response of a low-pass filter, we are then able to quickly determine the response of the corresponding high-pass filter, simply by using the 1/F transformation. Of course, we can do the reverse of predicting the low-pass response from the high-pass response.

CHAPTER 5

2nd Order VCVS Filters

INTRODUCTION

In this chapter, we will begin the discussion of the design of 2nd order low-pass and high-pass active filters by considering the voltage-controlled-voltage-source, or VCVS filters. Since the 2nd order filter has a stopband response that is twice as good as the 1st order filter, it is an important building block for the design of higher order filters.

OBJECTIVES

At the completion of this chapter, you will be able to do the following:

- Design and build both unity gain and "equal component" VCVS low-pass and high-pass filters
- Use scaling techniques

THE VCVS LOW-PASS FILTER

The simplest 2nd order low-pass filter is the voltage-controlled-voltage-source, or VCVS circuit of Fig. 5-1, which is also referred to as the *Sallen and Key filter*. For this circuit, the cutoff frequency is given by

$$\omega_c = \frac{1}{(R_1R_2C_3C_4)^{1/2}} \qquad \text{(Eq. 5-1)}$$

or

$$f_c = \frac{1}{2\pi(R_1R_2C_3C_4)^{1/2}} \qquad \text{(Eq. 5-2)}$$

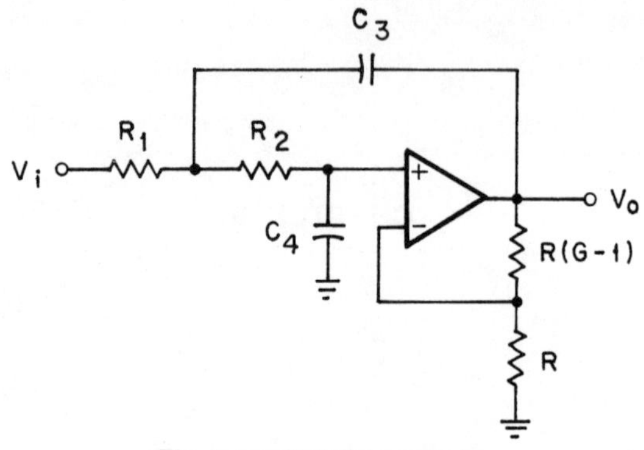

Fig. 5-1. VCVS filter circuit.

For computing purposes, the frequency response of any 2nd order low-pass filter having a passband gain G, is

$$dB = 20\log_{10}G - 20\log_{10}[\omega^4 + (\alpha^2 - 2)\omega^2 + 1]^{1/2} \qquad \text{(Eq. 5-3)}$$

However, before we can graph the amplitude response, we must assign a suitable value for the parameter α, which is the *damping factor,* and controls the shape of the overall amplitude response. In this book, only the *Butterworth* response will be considered, being easily understood and is the best overall choice.

For a 2nd order Butterworth response, $\alpha = 1.414$, and Equation 5-3 becomes

$$dB = 20\log_{10}G - 20\log_{10}[\omega^4+1]^{1/2} \qquad \text{(Eq. 5-4)}$$

The amplitude response is shown in Fig. 5-2, which is easily characterized as a Butterworth filter by its *maximally flat* passband. In the stopband, the response is twice that of a 1st order filter. Consequently the rolloff is −12 dB/octave or −40 dB/decade.

The big decision now facing us is: how do we pick the values for the resistors and capacitors for the VCVS circuit of Fig. 5-1? Since there are a number of possible combinations, we will discuss two of these possibilities:

- A VCVS filter with unity gain
- An "equal-component" VCVS filter

If the passband gain is unity and if $R_1 = R_2$, then $C_3 = 2C_4$, using a voltage follower configuration. For a cutoff frequency of 1 radian/s,

$$R_1 = R_2 = 1\Omega$$
$$C_3 = 1.414F$$
$$C_4 = 0.707F$$

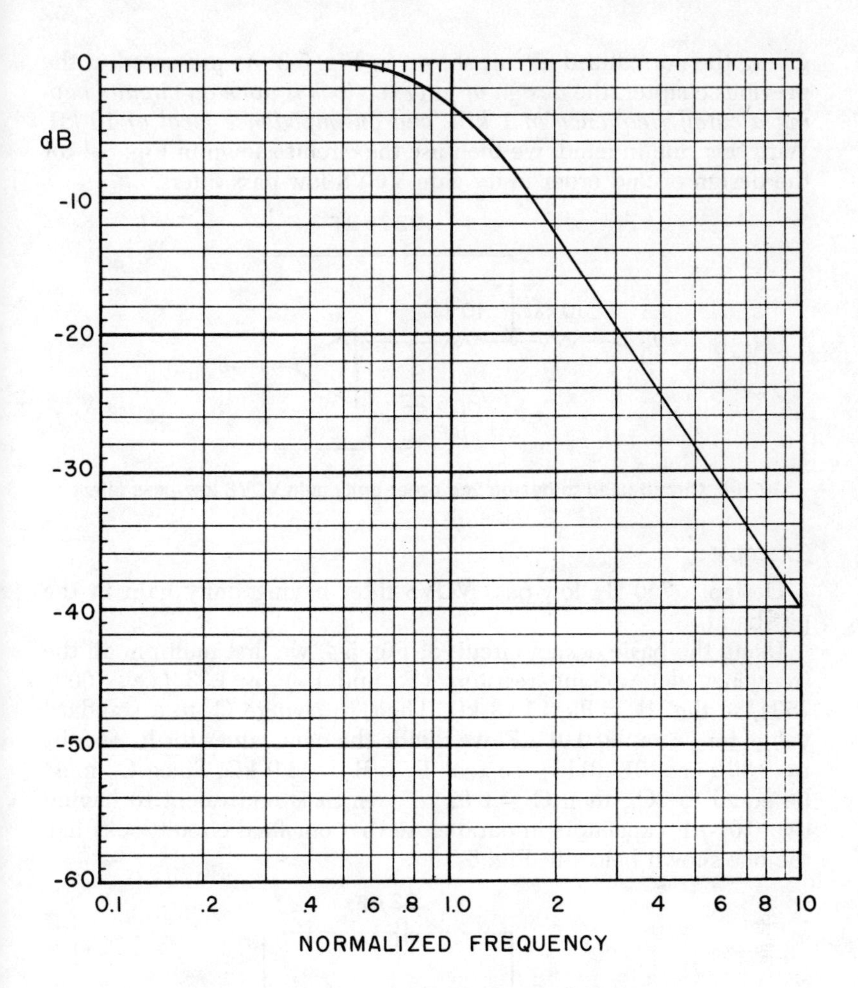

Fig. 5-2. Amplitude response of a Butterworth filter.

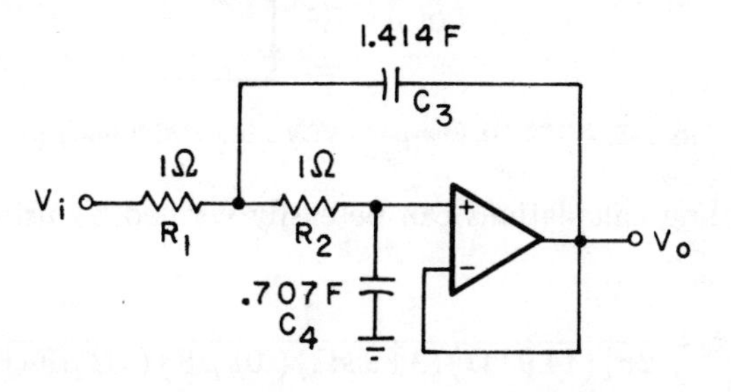

Fig. 5-3. Normalized circuit of Fig. 5-1.

giving the normalized circuit shown in Fig. 5-3. As presented in the previous chapter, the *design of filter sections is done on circuits having a cutoff frequency of 1 kHz and an impedance level of 10 kΩ.* With this rule in mind, we then use the circuit shown in Fig. 5-4 for the design of 2nd order unity gain VCVS low-pass filters.

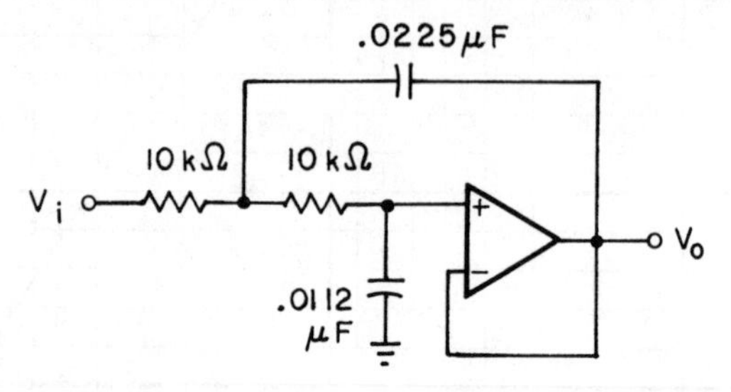

Fig. 5-4. Circuit used to design 2nd order unity gain VCVS low-pass filters.

Example:

Design a 750-Hz low-pass VCVS filter having unity gain in the passband.

Using the basic design circuit of Fig. 5-4, we first multiply all the frequency determining resistors (R_1 and R_2) by 1.33 (i.e., 1000/750), so that $R_1 = R_2 = 13.3$ kΩ. Then, to change C_4 to a standard value, for example, 0.01 μF, we divide the new values for R_1 and R_2 by 0.89 (i.e., .01/.0112), so that $R_1 = R_2 = 14.9$ kΩ. Since C_3 must be equal to $2C_4$, then $C_3 = 0.02$ μF, which is equivalent to having two 0.01-μF capacitors in parallel, so that our final circuit looks like the one shown below in Fig. 5-5.

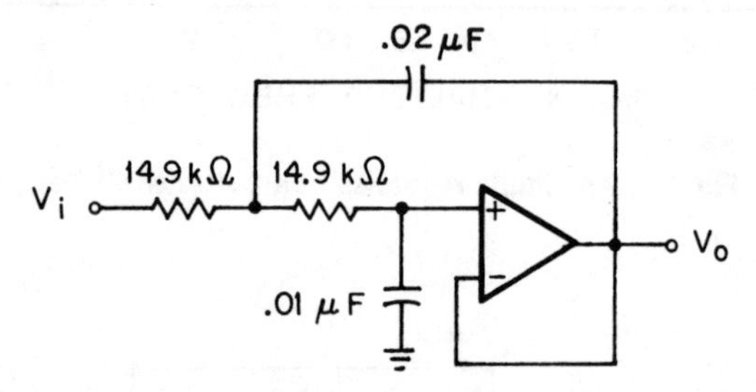

Fig. 5-5. A 750-Hz low-pass VCVS filter with unity gain.

Our scaling calculations can be easily verified by using Equation 5-2, or

$$f_c = \frac{1}{2\pi[(14.9\ k\Omega)(14.9\ k\Omega)(.01\ \mu F)(.02\ \mu F)]^{1/2}}$$
$$\simeq 750\ Hz$$

Alternatively, if $R_1 = R_2$, and $C_3 = C_4$, we have an "equal-component" VCVS low-pass filter, as shown in Fig. 5-6, for a normalized cutoff frequency of $\omega_c = 1$ radian/s. However, we pay a premium for

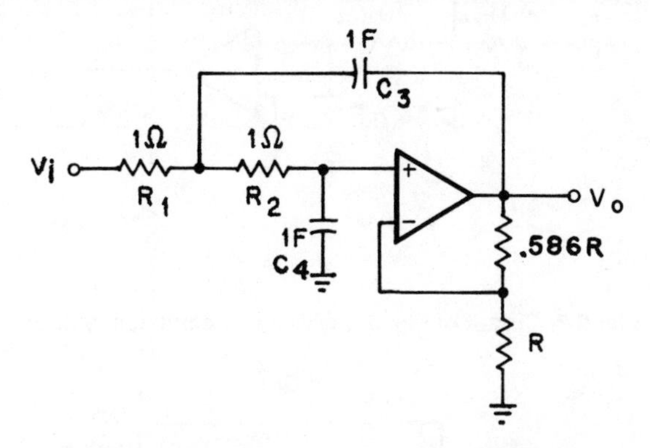

Fig. 5-6. Equal component VCVS low-pass filter.

the convenience of having equal resistors and capacitors. The passband gain will be fixed at a value equal to $G = 3 - \alpha$, or $G = 1.586$, since $\alpha = 1.414$ for a 2nd order Butterworth response. This is the *only gain that will permit the circuit to work.* The cutoff frequency will be at the point where the amplitude response is 3 dB less than the passband gain of 4 dB, or +1 dB. For an op amp in the *noninverting* mode, the feedback resistor must be 0.586 times the input resistor. Using ±5% resistors, a good choice for R_A and R_B is 47 kΩ and 27 kΩ respectively, as shown in Fig. 5-7. If ±1% resistors are available, $R_A = 51.1$ kΩ and $R_B = 30.1$ kΩ.

Example:

Design a 2 kHz "equal-component" low-pass filter.

Using Fig. 5-7, we first multiply R_1 and R_2 by 0.5 (i.e., 1000/2000) to give $R_1 = R_2 = 5$ kΩ. To change C_3 and C_4 to a standard value, for example .01 μF, we then divide the intermediate resistor values by 0.629 (i.e., .01/.0159), or $R_1 = R_2 = 7.95$ kΩ. The final circuit is shown in Fig. 5-8.

The main advantage of the "equal-component" VCVS filter over the unity gain VCVS filter is that both resistors and both capacitors are equal, thus simplifying the selection process. Since $R_1 = R_2$, and $C_3 = C_4$, Equation 5-2 reduces to

$$f_c = \frac{1}{2\pi R_1 C_3} \qquad \text{(Eq. 5-5)}$$

which is the same expression for 1st order filters! Consequently, we can use the graph of Fig. 4-8, which is given again in Fig. 5-9, to

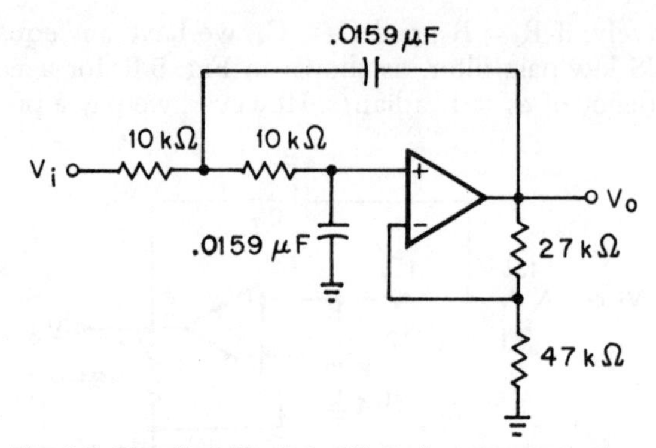

Fig. 5-7. Circuit of Fig. 5-6 with final component values.

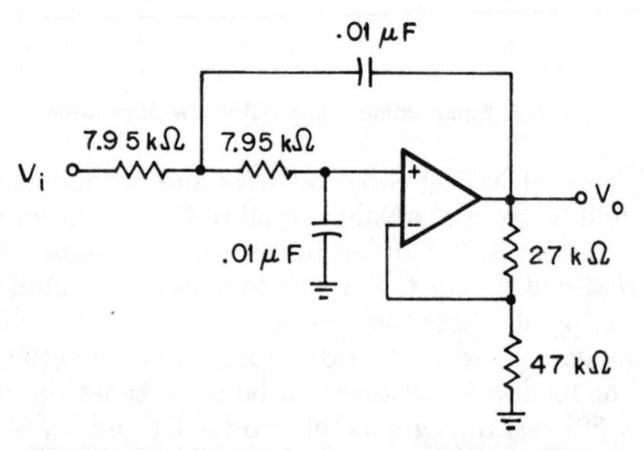

Fig. 5-8. A 2-kHz equal-component low-pass filter.

quickly determine the necessary resistor and capacitor values. *However, the graph of Fig. 5-9 cannot be used for the design of unity gain VCVS filters!*

THE VCVS HIGH-PASS FILTER

As with the 1st order filter, by simply interchanging the frequency-dependent resistors (R_1 and R_2) with the frequency-dependent capacitors (C_3 and C_4) of Fig. 5-1, we have now a 2nd order VCVS high-pass filter, shown in Fig. 5-10. For this circuit, the cutoff frequency is given by

$$\omega_c = \frac{1}{(R_3R_4C_1C_2)^{1/2}} \qquad \text{(Eq. 5-6)}$$

or

$$f_c = \frac{1}{2\pi(R_3R_4C_1C_2)^{1/2}} \qquad \text{(Eq. 5-7)}$$

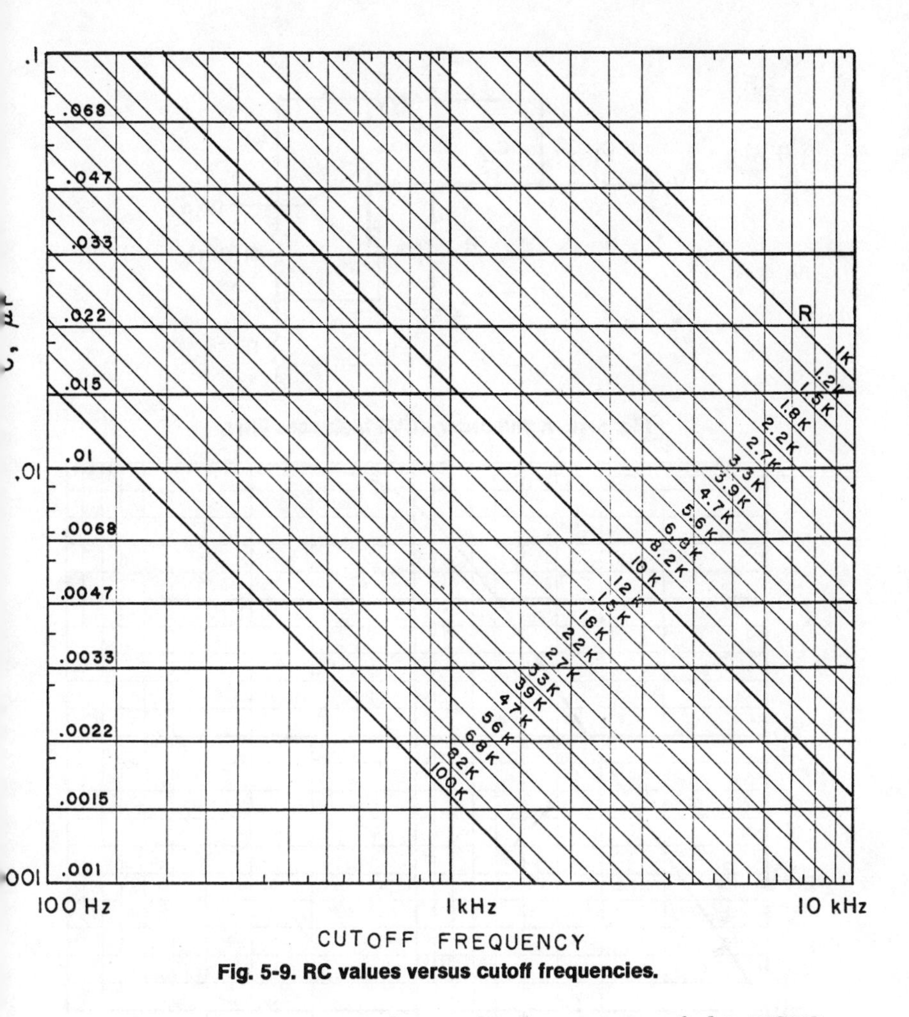

Fig. 5-9. RC values versus cutoff frequencies.

For computing purposes, the amplitude response of this VCVS high-pass filter is given by

$$dB = 20 \log_{10}G - 20 \log_{10}\left[(1/\omega)^4 + \frac{\alpha^2 - 2}{\omega^2} + 1\right]^{1/2} \qquad \text{(Eq. 5-8)}$$

and, since $\alpha = 1.414$ for a 2nd order Butterworth response, Equation 5-8 becomes

$$dB = 20 \log_{10}G - 20 \log_{10}[(1/\omega)^4 + 1]^{1/2} \qquad \text{(Eq. 5-9)}$$

and is plotted in Fig. 5-11 for a cutoff frequency normalized at $\omega/\omega_c = 1$. In the stopband, the rolloff rate is +12 dB/octave, or +40 dB/decade.

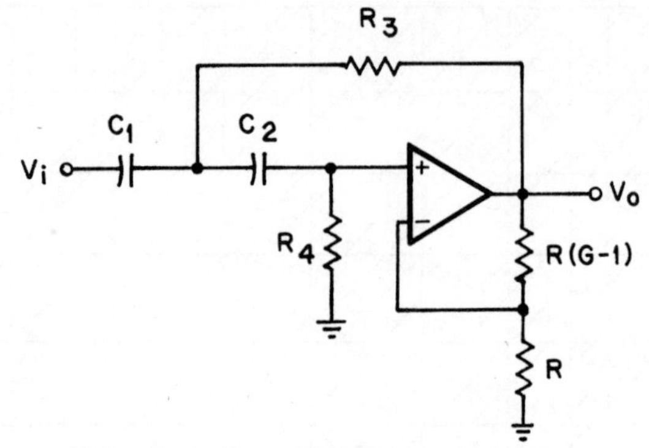

Fig. 5-10. A 2nd order VCVS high-pass filter.

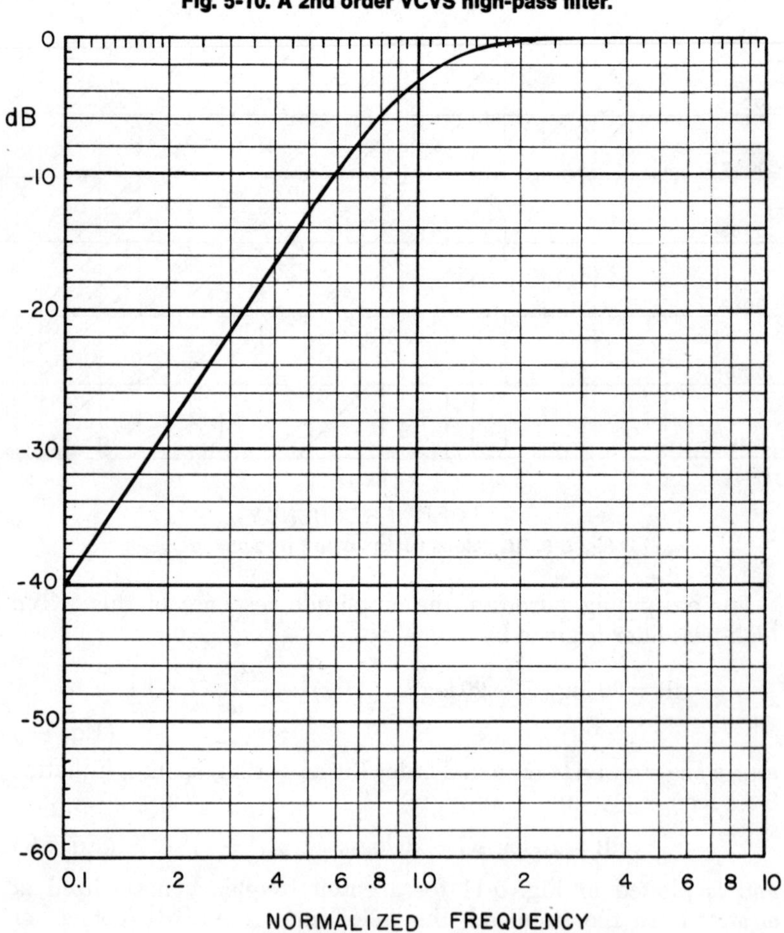

Fig. 5.11. Amplitude response of filter of Fig. 5-10.

As before, we can have several possibilities for the selection of Rs and Cs. For a passband gain of unity (0 dB), the easiest choice is to let $C_1 = C_2$, so that $R_4 = 2R_3$. For a cutoff frequency of 1 radian/s,

$$C_1 = C_2 = 1\text{F}$$
$$R_3 = 0.707\Omega$$
$$R_4 = 1.414\Omega$$

giving the normalized circuit of Fig. 5-12. In Fig. 5-13, we have the same circuit, but normalized for a cutoff frequency of 1 kHz, and an impedance level of 10 kΩ.

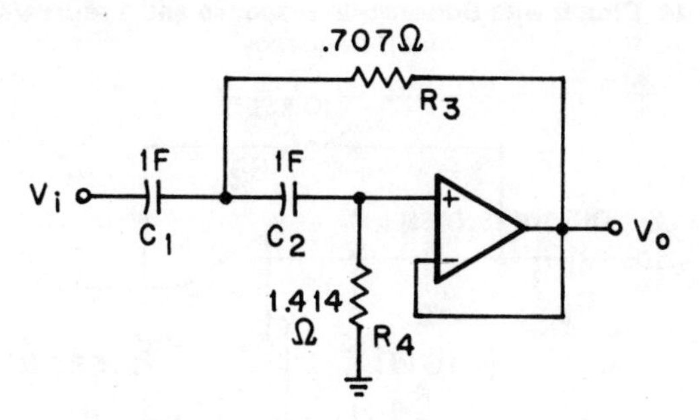

Fig. 5-12. Normalized circuit of Fig. 5-10.

Like the low-pass VCVS filter, there is also a similar "equal-component" high-pass VCVS filter, where $R_3 = R_4$, and $C_1 = C_2$, and the gain is fixed at $3 - \alpha$, or 1.586 for this type of circuit to have a Butterworth response. Fig. 5-14 has a cutoff frequency of 1 radian/s, while Fig. 5-15 is for a cutoff frequency of 1 kHz.

AN INTRODUCTION TO THE EXPERIMENTS

The following two experiments are designed to demonstrate the design and operation of the 2nd order VCVS (Sallen and Key) filter.

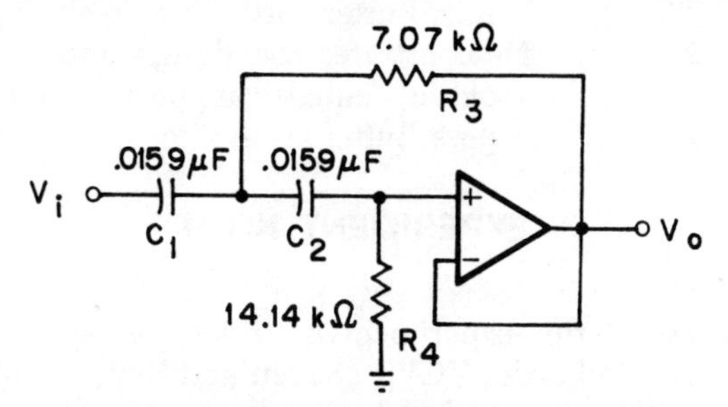

Fig. 5-13. Same circuit but normalized for cutoff at 1 kHz.

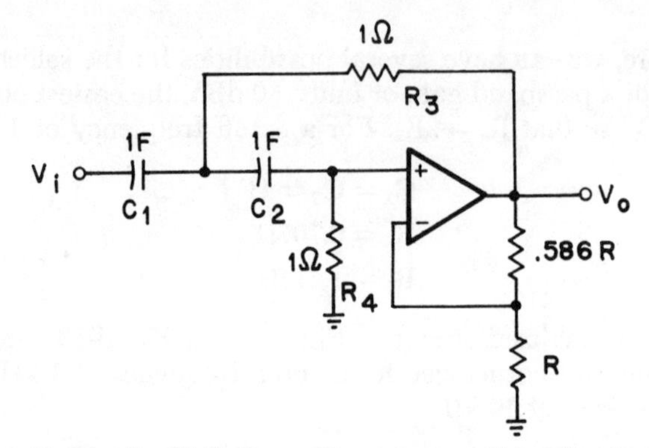

Fig. 5-14. Circuit with Butterworth response and 1 radian/second cutoff frequency.

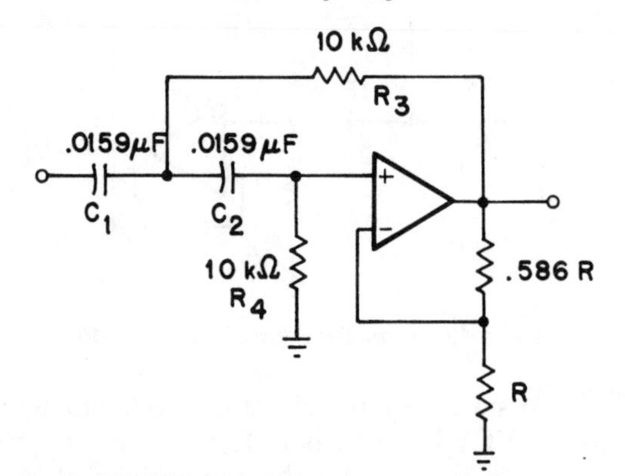

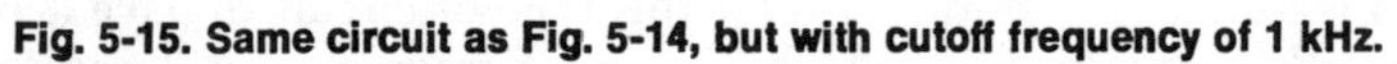

Fig. 5-15. Same circuit as Fig. 5-14, but with cutoff frequency of 1 kHz.

The experiments that you will perform can be summarized as follows:

Experiment No.	*Purpose*
1	Demonstrates the design and operation of a unity gain 2nd order VCVS low-pass Butterworth filter
2	Demonstrates the design and operation of the "equal component" VCVS low-pass Butterworth filter

EXPERIMENT NO. 1

Purpose

The purpose of this experiment is to demonstrate the operation and design of a 2nd order VCVS (Sallen and Key) low-pass Butterworth filter with unity passband gain.

Schematic Diagram of Circuit (Fig. 5-16)

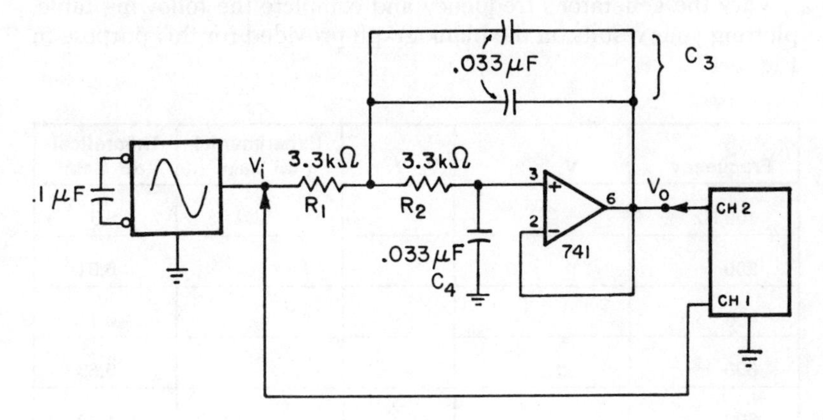

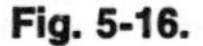

Fig. 5-16.

Design Basics

- Cutoff Frequency: $f_c = \dfrac{1}{2\pi(R_1R_2C_3C_4)^{1/2}}$
  where: $R_1 = R_2$ and $C_3 = 2C_4$
  for unity passband gain
- Amplitude Response: $20 \log_{10}\left[\dfrac{1}{(1 + f^4)^{1/2}}\right]$

Step 1

Set your oscilloscope for the following settings:

- Channels 1 & 2: 200 mV/division
- Time base: 1 ms/division
- AC coupling

Step 2

Wire the circuit shown in the schematic diagram. Apply power to the breadboard and adjust the input voltage to 1 volt peak-to-peak and a frequency of 100 Hz.

Step 3

With the values shown, what do you expect the cutoff frequency to be?

The cutoff frequency will be 1033 Hz, remembering that the two .033-µF capacitors in parallel equal .066 µF. For the purposes of this experiment, we will consider the cutoff frequency to be *1000 Hz.*

Step 4

Vary the generator's frequency and complete the following table, plotting your results on the blank graph provided for this purpose in Fig. 5-17.

Frequency	V_o	V_o/V_i	Experimental dB Gain	Theoretical dB Gain*
100 Hz				≃0
200				−0.01
400				−0.11
600				−0.53
800				−1.49
1000				−3.01
2000				−12.3
4000				−24.1
8000				−36.1
10,000				−40.0

* Based on a cutoff frequency of 1 kHz, for simplicity.

Step 5

As a comparison of the 1st and 2nd order unity gain low-pass filters' responses, plot the results of the next-to-last column of Step 4 of Experiment No. 1 also on this same graph. How do these two curves differ?

The 2nd order curve of this experiment has a stopband rolloff that is *twice* that of the 1st order curve, noting that they both should intersect at −3.0 dB when the frequency is 1 kHz. Beyond a frequency of 2 kHz, what is the rolloff (i.e, the rate of decrease) of the response of the 2nd order filter that you have just determined experimentally?

Your experimental result should approximately equal −12 dB/octave, or −40 dB/decade. Two convenient frequencies that you have already determined the filter's response is 4 kHz and 8 kHz, which are 1 octave apart. Consequently, the difference in the response between these two points should almost be −12 dB. From the last column in the table of Step 4, the decrease is 24.1 dB − 36.1 dB, or −12 dB.

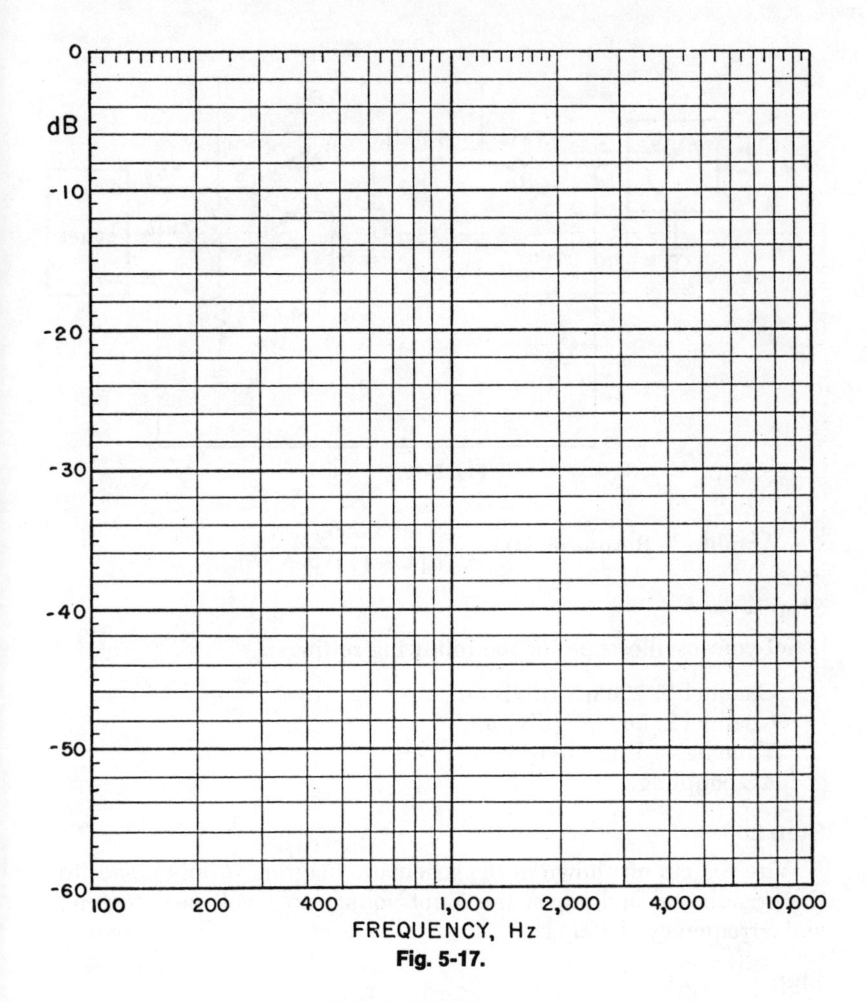

Fig. 5-17.

EXPERIMENT NO. 2

Purpose

The purpose of this experiment is to demonstrate the operation and design of the "equal component" VCVS low-pass filter.

Schematic Diagram of Circuit (Fig. 5-18)

Design Basics

- Cutoff Frequency: $f_c = \dfrac{1}{2\pi(R_1R_2C_3C_4)^{1/2}}$
  where: $R_1 = R_2$ and $C_3 = C_4$
- Gain: fixed at $3 - \alpha = 1.586$ for Butterworth response where α must equal 1.414.

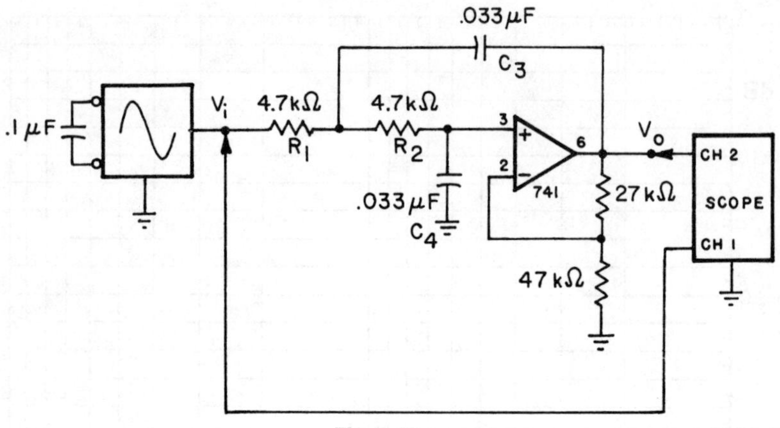

Fig. 5-18.

- Amplitude Response: $20 \log_{10}\left[\frac{1.586}{[1+(f)^4]^{1/2}}\right]$

Step 1

Set your oscilloscope for the following settings:

- Channel 1: 200 mV/division
- Channel 2: 500 mV/division
- Time base: 1 ms/division
- AC coupling

Step 2

Wire the circuit shown in the schematic diagram. Apply power to the breadboard and adjust the input voltage to 1 volt peak-to-peak and a frequency of 100 Hz.

Step 3

Vary the generator frequency and complete the following table, plotting your results on the blank graph provided for this purpose in Fig. 5-19. To complete the last column, *subtract the gain measured at 100 Hz (i.e., the passband gain) from the dB gain for each measured frequency.* Plot the results of the *next-to-the-last column.*

Frequency	V_o	V_o/V_i	dB Gain	Normalized dB Gain
100 Hz				
200				
400				

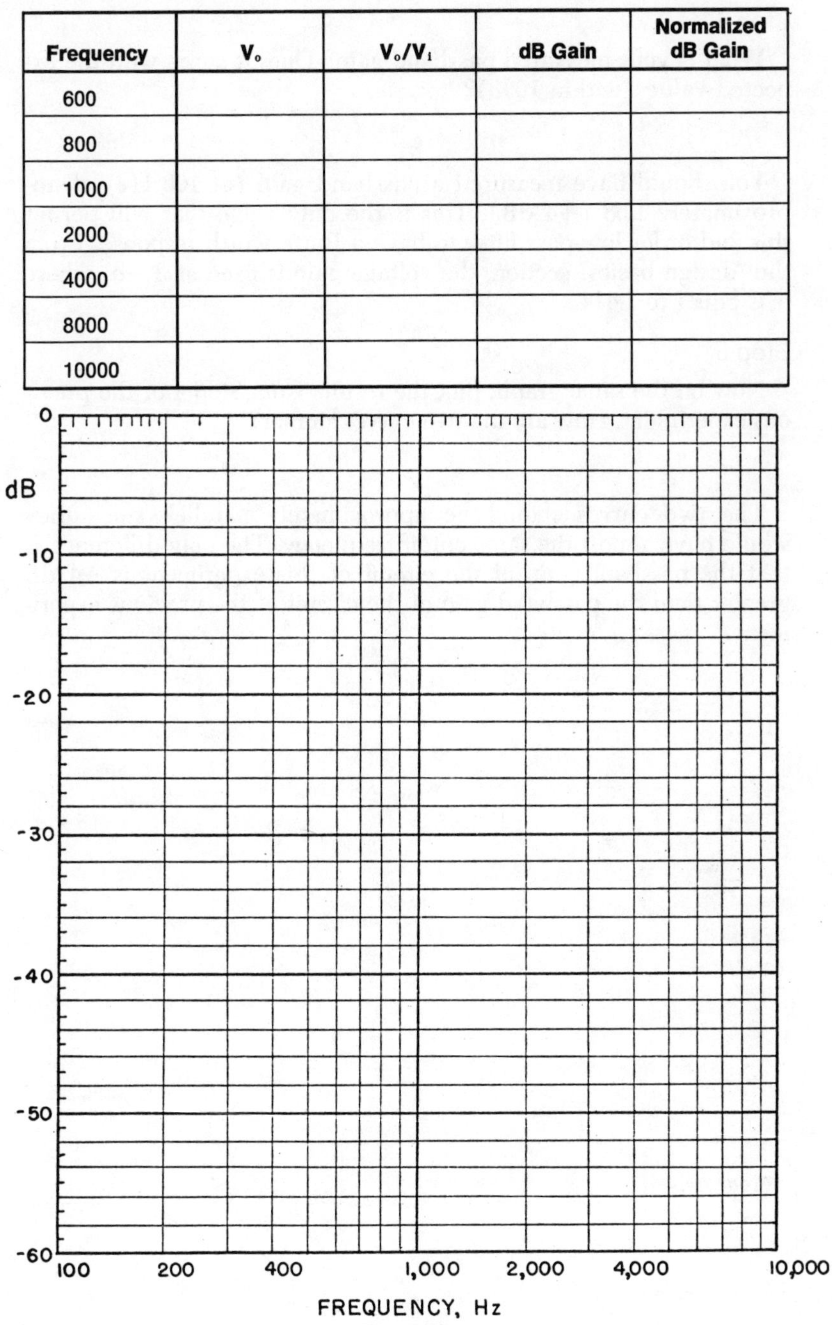

Frequency	V_o	V_o/V_i	dB Gain	Normalized dB Gain
600				
800				
1000				
2000				
4000				
8000				
10000				

Fig. 5-19.

Step 4

What is your measured passband gain? Does it agree with the expected value (within 10%)?

You should have measured a passband gain (at 100 Hz) of approximately 1.58 (+4 dB). This is the only value that will permit this 2nd order low-pass filter to have a Butterworth response. From the "design basics" section, the voltage gain is fixed at $3 - \alpha$, where α is equal to 1.414.

Step 5

Now on the same graph, plot the results from Step 4 of the previous experiment. How are the two curves similar?

The two curves should be approximately parallel, since they would have almost the same cutoff frequency. The only difference is that the passband gain of the circuit of this experiment is +4 dB greater than the passband gain of the circuit of the previous experiment.

CHAPTER 6

2nd Order Multiple-Feedback Filters

INTRODUCTION

In this chapter another 2nd order filter will be described, having the same normalized amplitude responses as the VCVS (Sallen and Key) filter. However, in actual practice, the VCVS filter is preferred, especially for the formation of the higher order filters of the next chapter. On the other hand, the multiple-feedback network is important, enjoying widespread use in the design of bandpass filters (Chapter 8).

OBJECTIVES

At the completion of this chapter, you will be able to do the following:

- Learn how to design and build 2nd order low-pass and high-pass multiple-feedback filters
- Compare the design of VCVS filters with multiple-feedback filters

THE MULTIPLE-FEEDBACK LOW-PASS FILTER

The 2nd order multiple-feedback low-pass filter circuit is shown in Fig. 6-1. There are two obvious differences between this circuit and the VCVS filter:

- There is an additional feedback path, hence the name, *multiple-feedback filter.*
- The operational amplifier is connected in the *inverting* mode.

For this circuit, the cutoff frequency is determined by

$$\omega_c = \frac{1}{(R_2R_4C_3C_5)^{1/2}} \qquad \text{(Eq. 6-1)}$$

or

$$f_c = \frac{1}{2\pi(R_2R_4C_3C_5)^{1/2}} \qquad \text{(Eq. 6-2)}$$

For computing the amplitude response, the equation is the same as Equation 5-3, which reduces to Equation 5-4, or

$$dB = 20 \log_{10}G - 20 \log_{10}[\omega^4 + 1]^{1/2} \qquad \text{(Eq. 6-3)}$$

for a 2nd order Butterworth response. The passband voltage gain G is determined solely by the ratio of R_1 and R_4, so that

$$G = \frac{R_4}{R_1} \qquad \text{(Eq. 6-4)}$$

We now have one more frequency-determining component than the VCVS filter. Consequently, the determination of the component values for the circuit of Fig. 6-1 becomes a bit complicated, as for the VCVS filters, unless some simplifications are made.

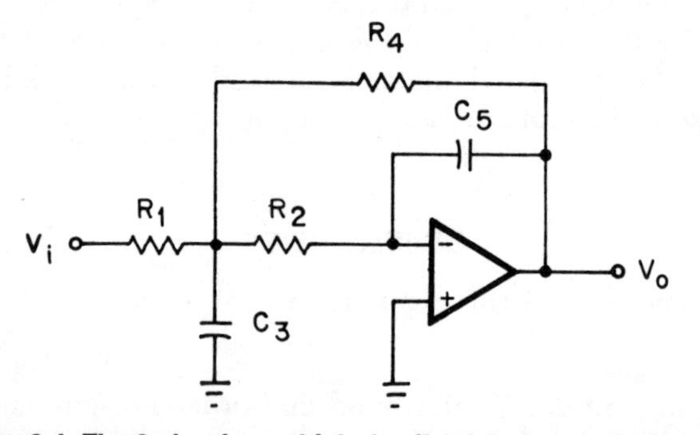

Fig. 6-1. The 2nd order multiple-feedback low-pass filter circuit.

If the passband voltage gain is to be unity and the cutoff frequency is 1 radian/s, then for this filter to have a 2nd order Butterworth response,

$$R_1 = R_2 = R_4 = 1\Omega$$
$$C_3 = 2.12F$$
$$C_5 = 0.471F$$

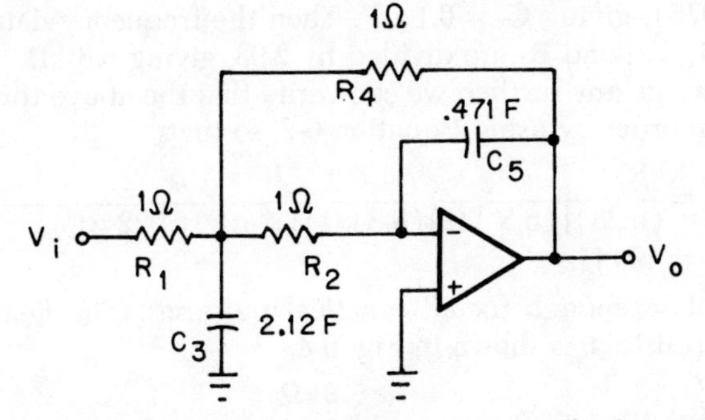

Fig. 6-2. Circuit with cutoff frequency of 1 radian/second.

giving the 1 radian/s-1Ω circuit of Fig. 6-2. Again, as in the previous two chapters, for *designing* filters, we are more at ease in using a filter circuit that has been normalized for a cutoff frequency of 1 kHz and an impedance level of 10 kΩ, as shown in Fig. 6-3.

Unless we remove all restrictions on the components of the network of Fig. 6-1, we will find that the values assigned to the basic circuit of Fig. 6-2 are the only values possible. Therefore, in order to have a Butterworth response, the passband gain is fixed at 1; we practically have no other choice!

Example:

Design a 500-Hz multiple-feedback low-pass filter with unity passband gain.

Since the cutoff frequency is 500 Hz, we multiply all frequency-determining resistors (i.e., R_2 and R_4) of Fig. 6-3 by 2.0 (i.e., 1000 Hz/500 Hz), so that $R_1 = R_2 = R_4 = 20$ kΩ. To change the frequency-determining capacitors to standard values, let the new value for capacitor C_5 be, for example, .022 μF. Then C_3 is multiplied by the ratio of the new value for C_5 by its original value, or 2.93 (i.e.,

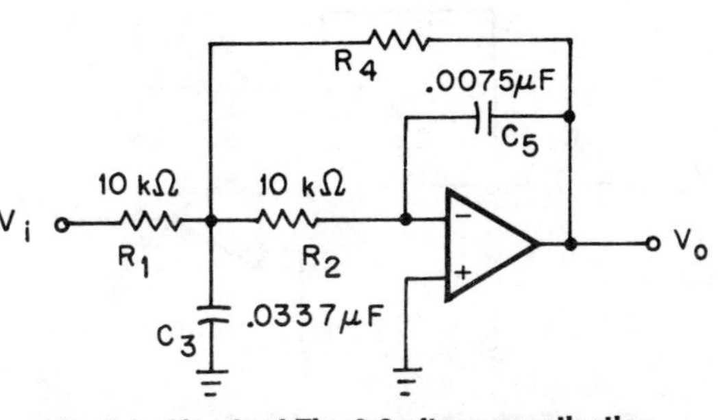

Fig. 6-3. Circuit of Fig. 6-2 after normalization.

0.022/0.0075), giving $C_3 = 0.1\ \mu F$. Then the frequency-determining resistors R_1, R_2, and R_4 are divided by 2.93, giving 6.8 kΩ.

Before we go any further, we can verify that the above transformations are in order by using Equation 6-2, so that

$$f_c = \frac{1}{(6.28)[(6.8\ k\Omega)(6.8\ k\Omega)(.1\ \mu F)(.022\ \mu F)]^{1/2}}$$
$$= 499\ Hz$$

which is close enough for all practical purposes. The final circuit, using 5% resistors, is shown in Fig. 6-4.

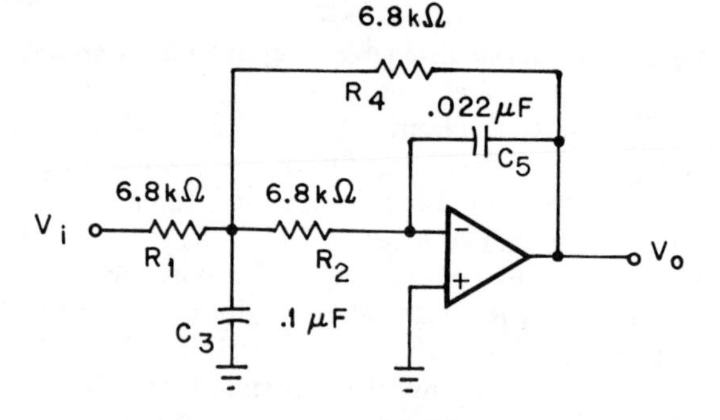

Fig. 6-4. A 500-Hz multiple-feedback low-pass filter with unity passband gain.

THE MULTIPLE-FEEDBACK HIGH-PASS FILTER

In the previous two chapters, the corresponding high-pass filter was formed by simply interchanging the position of the frequency-determining resistors with capacitors, and vice versa. The same applies to the formation of the multiple-feedback high-pass filter of Fig. 6-5. In terms of the circuit parameters, the cutoff frequency is either

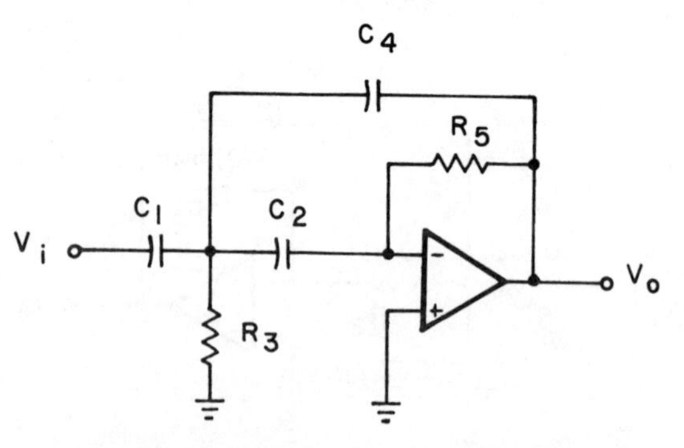

Fig. 6-5. Multiple-feedback high-pass filter.

$$\omega_c = \frac{1}{(R_3R_5C_2C_4)^{1/2}} \qquad \text{(Eq. 6-5)}$$

or

$$f_c = \frac{1}{2\pi(R_3R_5C_2C_4)^{1/2}} \qquad \text{(Eq. 6-6)}$$

The amplitude response is the same as Equation 5-9,

$$dB = 20 \log_{10}G - 20 \log_{10}[(1/\omega)^4 + 1]^{1/2} \qquad \text{(Eq. 6-7)}$$

for a 2nd order Butterworth response.

This high-pass circuit is different from all other op-amp circuits in that *the passband gain is set by the ratio of two capacitors, instead of resistors,* so that

$$G = \frac{C_4}{C_1} \qquad \text{(Eq. 6-8)}$$

Is seems that in order to simplify the selection of the circuit components, we wind up making compromises, and this circuit is no exception. Consequently, the only "reasonable" compromise is to have a *unity gain multiple-feedback filter,* so that the three capacitors are equal. For a cutoff frequency of 1 radian/s, we have the component values shown in Fig. 6-6. However, for designing 2nd

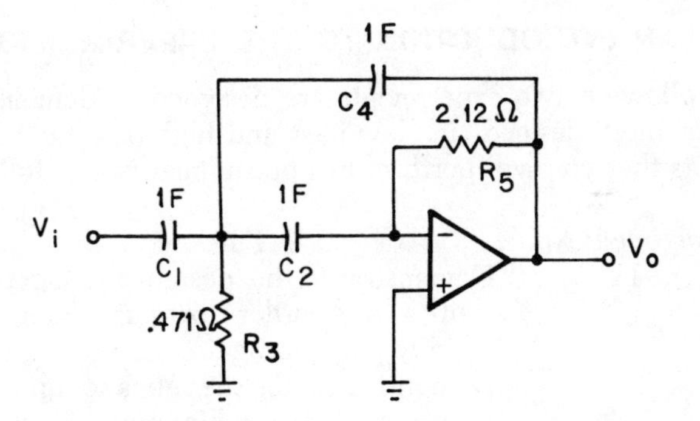

Fig. 6-6. Unity-gain multiple-feedback filter with a cutoff frequency of 1 radian/second.

order multiple-feedback high-pass filters, it is more convenient to use the unity gain 1-kHz circuit of Fig. 6-7.

SOME COMMENTS

The design of the low-pass and high-pass multiple-feedback filters is a bit more complicated than the 2nd order VCVS filters of Chapter

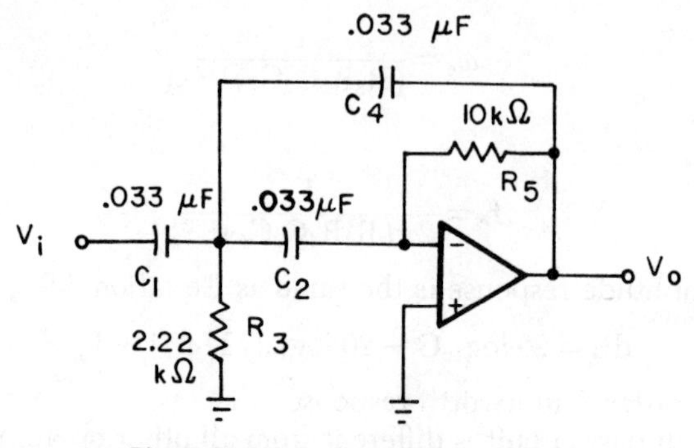

Fig. 6-7. Unity-gain, 1-kHz circuit.

5. This is particularly true when it is desired to shift the filter's cutoff frequency easily without affecting the passband gain and the Butterworth response. This is the disadvantage of using the op amp in the inverting mode.

In practice, the VCVS filter is preferred for the construction of either 2nd order low-pass or high-pass filters, especially for the higher order filters of the next chapter. However, as will be discussed in Chapter 8, the basic multiple-feedback circuit configuration is the best choice for *bandpass filters.*

AN INTRODUCTION TO THE EXPERIMENTS

The following two experiments are designed to demonstrate the design of multiple-feedback low-pass and high-pass filters. The experiments that you will perform can be summarized as follows:

Experiment No.	*Purpose*
1	Demonstrates the design and operation of a 2nd order low-pass filter with unity passband gain.
2	Demonstrates the design and operation of a 2nd order high-pass filter with unity passband gain.

EXPERIMENT NO. 1

Purpose

The purpose of this experiment is to demonstrate the operation and design of a 2nd order multiple-feedback low-pass filter with unity passband gain and a Butterworth amplitude response.

Schematic Diagram of Circuit (Fig. 6-8)

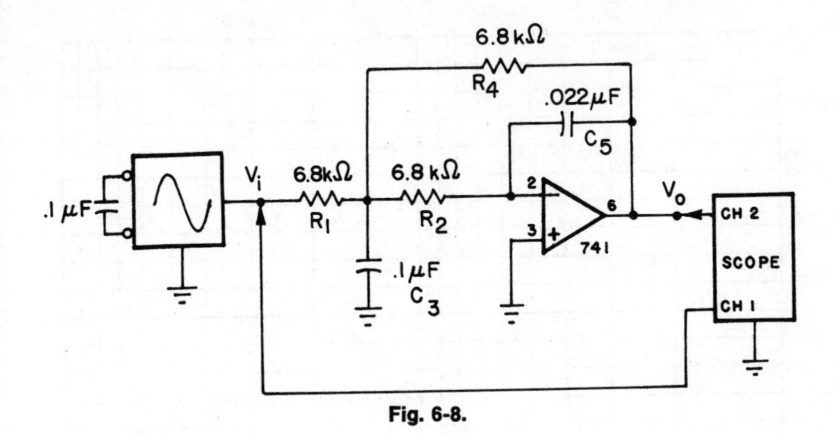

Fig. 6-8.

Design Basics

- Cutoff frequency: $f_c = \dfrac{1}{2\pi[R_2R_4C_3C_5]^{1/2}}$
  where $R_2 = R_4$ and $C_3 = 4.5\ C_5$ for a Butterworth response
- Passband gain: $G = \dfrac{R_4}{R_1} = 1.0$ for a Butterworth response
- Amplitude response: $-20\ \log_{10}[1 + (f)^4]^{1/2}$

Step 1

Set your oscilloscope for the following settings:

- Channels 1 & 2: 0.5 volt/division
- Time base: 1 ms/division
- AC coupling

Step 2

Wire the circuit in the schematic diagram. Apply power to the breadboard and adjust the input voltage at 2 volts peak-to-peak and a frequency of 100 Hz.

Step 3

Vary the generator's frequency and complete the following table, plotting your results on the blank graph provided for this purpose in Fig. 6-9. From either the table or your graph, what is the cutoff frequency? How does it compare with the expected value?

The expected frequency, according to the formula given in the "design basics" section of this experiment is 499 Hz. For simplicity, we'll take the cutoff frequency to be 500 Hz.

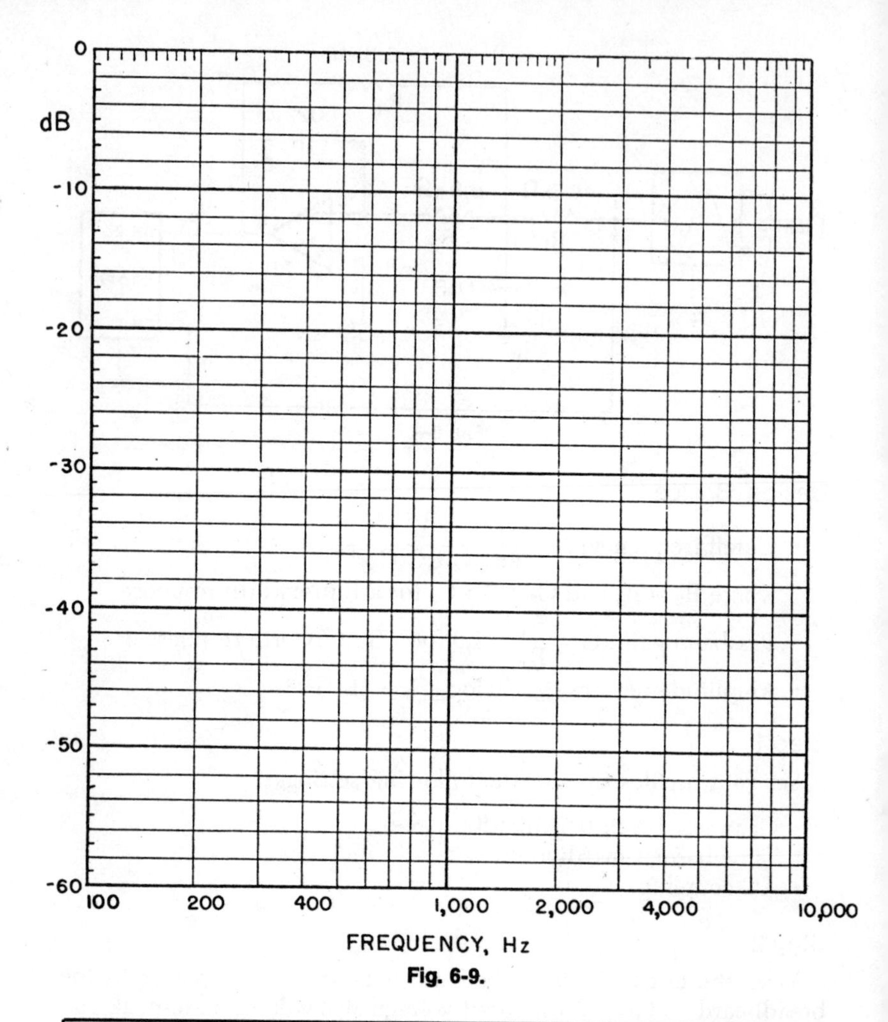

Fig. 6-9.

Frequency	V_o	V_o/V_1	Experimental dB Gain
100 Hz			
200			
300			
400			
500			
600			
700			

Frequency	V_o	V_o/V_1	Experimental dB Gain
800			
900			
1000			
2000			
3000			
4000			
5000			

Step 4

Subtract the dB gain you measured at 2000 Hz from the measured value at 4000 Hz (1 octave). Is this a 2nd order low-pass filter?

The amplitude response for this 1 octave should closely equal −12 dB, which is characteristic of a 2nd order low-pass response.

EXPERIMENT NO. 2

Purpose

The purpose of this experiment is to demonstrate the operation and design of a 2nd order multiple-feedback high-pass filter with unity passband gain and a Butterworth amplitude response.

Schematic Diagram of Circuit (Fig. 6-10)

Design Basics

- Cutoff frequency: $f_c = \dfrac{1}{2\pi(R_3R_5C_2C_4)^{1/2}}$
  where $C_2 = C_4$ and $R_5 = 4.5\ R_3$ for a Butterworth response
- Passband gain: $G = C_4/C_1 = 1.0$ for a Butterworth response
- Amplitude response: $-20\log_{10}[1 + (1/f)^4]^{1/2}$

Step 1

Set your oscilloscope for the following settings:

- Channel 1: 0.5 volt/division
- Channel 2: 10 mV/division
- Time base: 1 ms/division
- AC coupling

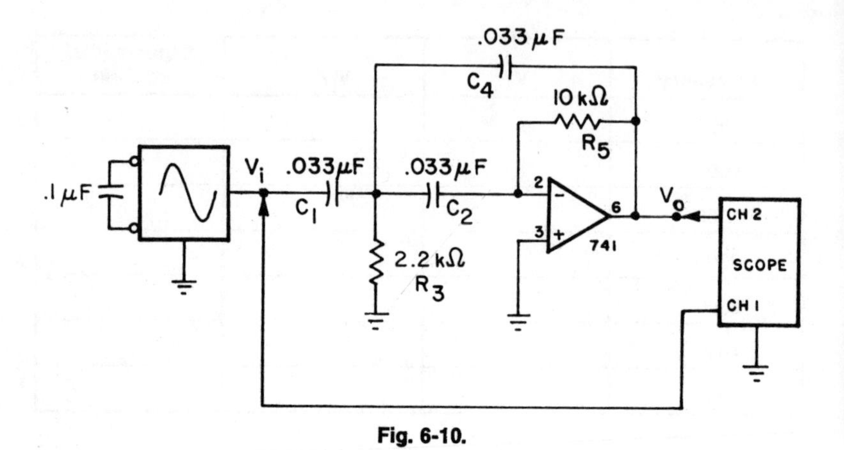

Fig. 6-10.

Step 2

Wire the circuit shown in the schematic diagram. Apply power to the breadboard and adjust the input voltage at 2 volts peak-to-peak and the frequency at 100 Hz.

Step 3

Vary the generator frequency and complete the following table, plotting your results on the blank graph provided for this purpose in Fig. 6-11. From either the table or the graph, what is the cutoff frequency? How does it compare with the expected value?

Frequency	V_o	V_o/V_i	Experimental dB Gain
100 Hz			
200			
300			
400			
500			
600			
700			
800			
900			
1000			

Frequency	V_o	V_o/V_i	Experimental dB Gain
2000			
3000			
4000			
5000			

The expected frequency, according to the formula given is 1028 Hz.

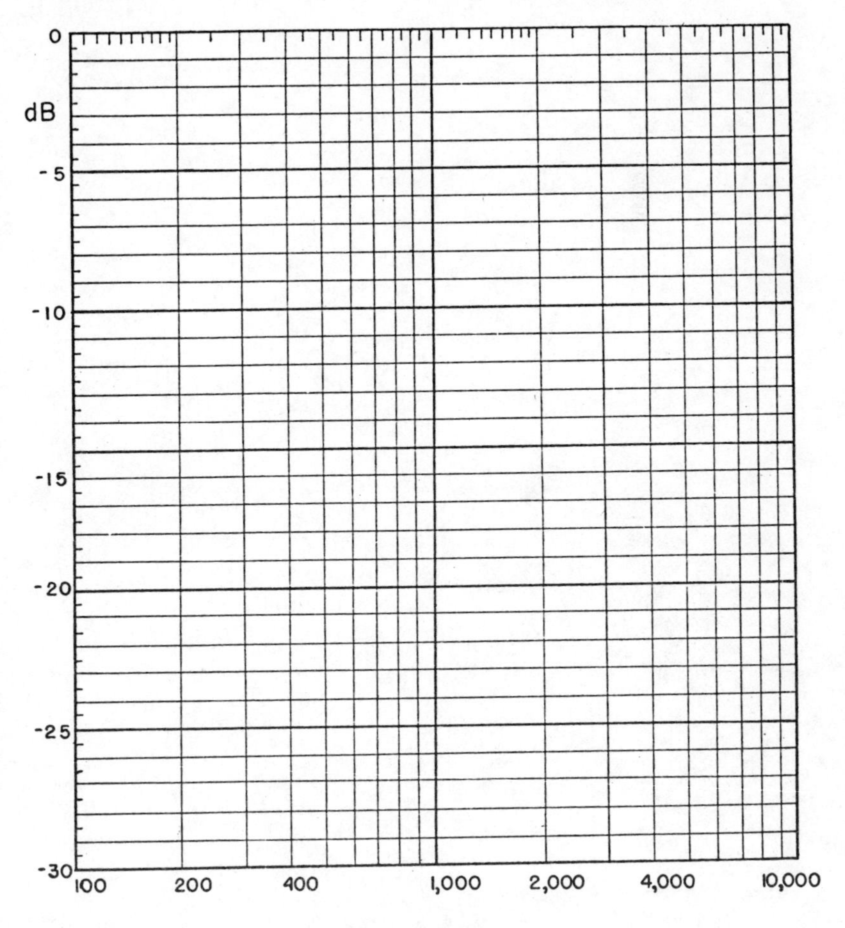

Fig. 6-11.

CHAPTER 7

Higher Order Filters

INTRODUCTION

In this chapter we will learn how to properly cascade 1st and 2nd order filter sections to form higher order low-pass and high-pass Butterworth filters.

OBJECTIVES

At the completion of this chapter, you will be able to do the following:

- Learn how to properly cascade filter sections
- Design and build higher order low-pass and high-pass Butterworth filters

CASCADING

Active filters with an order higher than 2 are formed by properly connecting in series, or *cascading*, 1st and 2nd order filter sections. As shown in Fig. 7-1, a 3rd order filter is formed by cascading a 1st and 2nd order section. A 4th order filter is formed by cascading two 2nd order sections, and so on. However, when there is more than one 2nd order filter section that is cascaded, *the 2nd order sections are not identical!*

To properly cascade the individual sections for a given order filter, the dampling factors for each filter section must equal those given in Table 7-1.*

1st order filters, like those presented in Chapter 4, *will always have a damping factor equal to 1.000.* Consequently we don't have

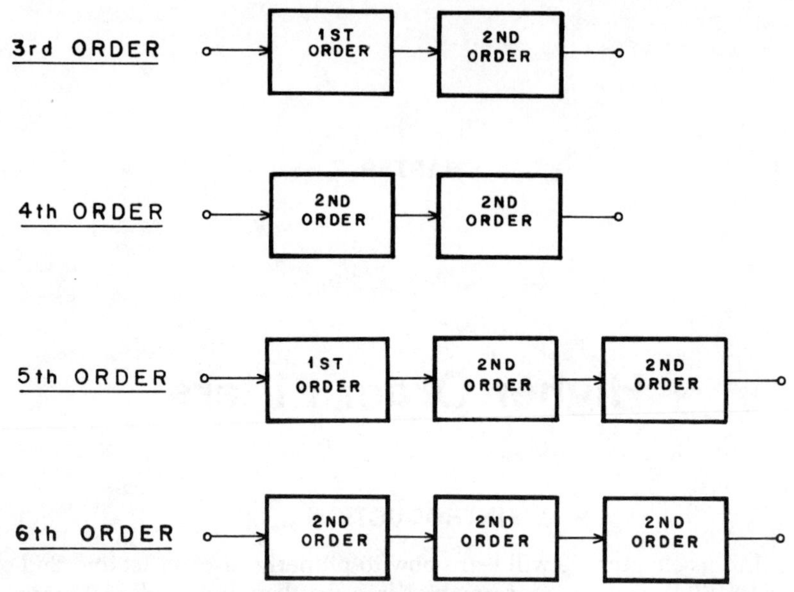

Fig. 7-1. Cascading filters.

Table 7-1. Damping Factors Used to Cascade Individual Filter Sections

Order	1st Section	2nd Section	3rd Section
1	1.000	—	—
2	1.414	—	—
3	1.000	1.000	—
4	1.848	0.765	—
5	1.000	1.618	0.618
6	1.932	1.414	0.518

* The derivation of these damping factors is given in Appendix C.

too much of a problem with the first section of a 3rd, or 5th order filter (or any other *odd order* filter). However, for a 2nd order section, the determination of the necessary component values to give the factors in Table 7-1 will be a little difficult, unless we restrict ourselves to using *only "equal-component" VCVS 2nd order filter sections.* Now, knowing what the proper damping factors are required for any order filter, half our problem is solved.

As pointed out in Chapter 5, the passband voltage gain and the damping factor of the "equal-component" filter are related to each other by the equation

$$G = 3 - \alpha \qquad \text{(Eq. 7-1)}$$

and since the passband voltage gain for this filter is simply the expression used for a noninverting amplifier

$$G = 1 + \frac{R_B}{R_A} \qquad \text{(Eq. 7-2)}$$

the values for R_A and R_B are then determined by combining Equations 7-1 and 7-2,

$$\frac{R_B}{R_A} = 2 - \alpha \qquad \text{(Eq. 7-3)}$$

Table 7-2 gives the best combinations of standard 5% resistor values to satisfy the requirements of Table 7-1 for the different 2nd order sections. If 1% resistors are available, Table 7-3 should be used.

When the individual sections are cascaded, the overall passband gain is now dependent on the gain of the individual sections, as shown in Table 7-4. In terms of voltage gain, *the overall voltage gain is the product of the individual voltage gains of the filter sections.* If it is desired to express the passband gain in terms of dB, *the overall dB gain is the sum of the individual dB gains of the filter sections.*

For example, a 5th order VCVS Butterworth filter (either low-pass or high-pass) will be constructed by cascading a single 1st order

Table 7-2. Standard 5% Resistor Values to Satisfy Requirements of Table 7-1

Order	1st Section	2nd Section	3rd Section
2	$R_A = 47\ k\Omega$ $R_B = 27\ k\Omega$	—	—
3	—	$R_A = 10\ k\Omega$ $R_B = 10\ k\Omega$	—
4	$R_A = 15\ k\Omega$ $R_B = 2.2\ k\Omega$	$R_A = 18\ k\Omega$ $R_B = 22\ k\Omega$	—
5	—	$R_A = 39\ k\Omega$ $R_B = 15\ k\Omega$	$R_A = 12\ k\Omega$ $R_B = 18\ k\Omega$
6	$R_A = 100\ k\Omega$ $R_B = 6.8\ k\Omega$	$R_A = 47\ k\Omega$ $R_B = 27\ k\Omega$	$R_A = 15\ k\Omega$ $R_B = 22\ k\Omega$

section with two 2nd order sections. The voltage gain of the 1st order section is 1.000 (0 dB), and the voltage gain of the 2nd order sections are 1.382 (2.8 dB) and 2.382 (7.5 dB) respectively, as shown in Table 7-4. Then, the overall voltage gain for our 5th order filter is

Table 7-3. Standard 1% Resistor Values to Satisfy Requirements of Table 7-1

Order	1st Section	2nd Section	3rd Section
2	$R_A = 17.4\ k\Omega$ $R_B = 10.2\ k\Omega$	—	—
3	—	$R_A = 10\ k\Omega$ $R_B = 10\ k\Omega$	—
4	$R_A = 15\ k\Omega$ $R_B = 2.26\ k\Omega$	$R_A = 10.7\ k\Omega$ $R_B = 13.7\ k\Omega$	—
5	—	$R_A = 39.2\ k\Omega$ $R_B = 15\ k\Omega$	$R_A = 10.7\ k\Omega$ $R_B = 14.7\ k\Omega$
6	$R_A = 100\ k\Omega$ $R_B = 6.81\ k\Omega$	$R_A = 17.4\ k\Omega$ $R_B = 10.2\ k\Omega$	$R_A = 13.3\ k\Omega$ $R_B = 19.6\ k\Omega$

$$\begin{aligned}\text{Gain} &= (1.000) \times (1.382) \times (2.382)\\ &= 3.29\end{aligned}$$

or, in terms of dB,

$$\begin{aligned}\text{Gain} &= (0\ \text{dB}) + (2.8\ \text{dB}) + (7.5\ \text{dB})\\ &= 10.3\ \text{dB}\end{aligned}$$

Although it is theoretically possible to construct higher order filters by cascading the necessary filter sections in any order we like, we will cascade the sections in the order of *decreasing damping,* as shown in Table 7-1 and Fig. 7-1. This is the same as cascading the sections in *increasing voltage gain,* as shown in Table 7-4.

Table 7-4. Overall Passband Gain Is Dependent on Indivdual Sections Gain

Order	1st Section	2nd Section	3rd Section	Overall
2	1.586 (+4.0 dB)	—	—	1.59 (+4.0 dB)
3	1.000 (0 dB)	2.000 (+6.0 dB)	—	2.00 (+6.0 dB)
4	1.152 (+1.2 dB)	2.235 (+7.0 dB)	—	2.57 (+8.2 dB)
5	1.000 (0 dB)	1.382 (+2.8 dB)	2.382 (+7.5 dB)	3.29 (+10.3 dB)
6	1.068 (+0.6 dB)	1.586 (+4.0 dB)	2.482 (+7.9 dB)	4.20 (+12.5 dB)

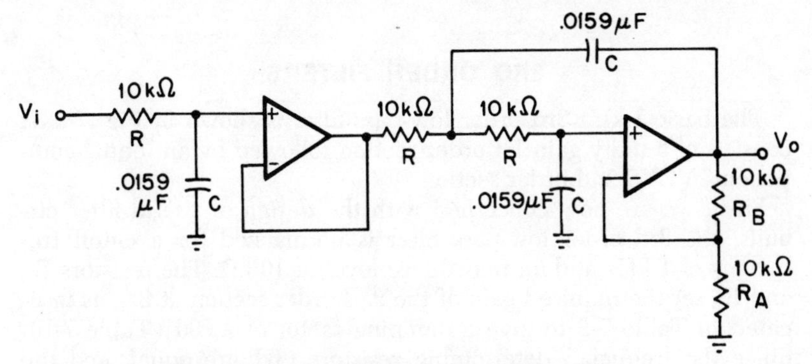

Fig. 7-2. Basic 1-kHz 3rd order low-pass filter.

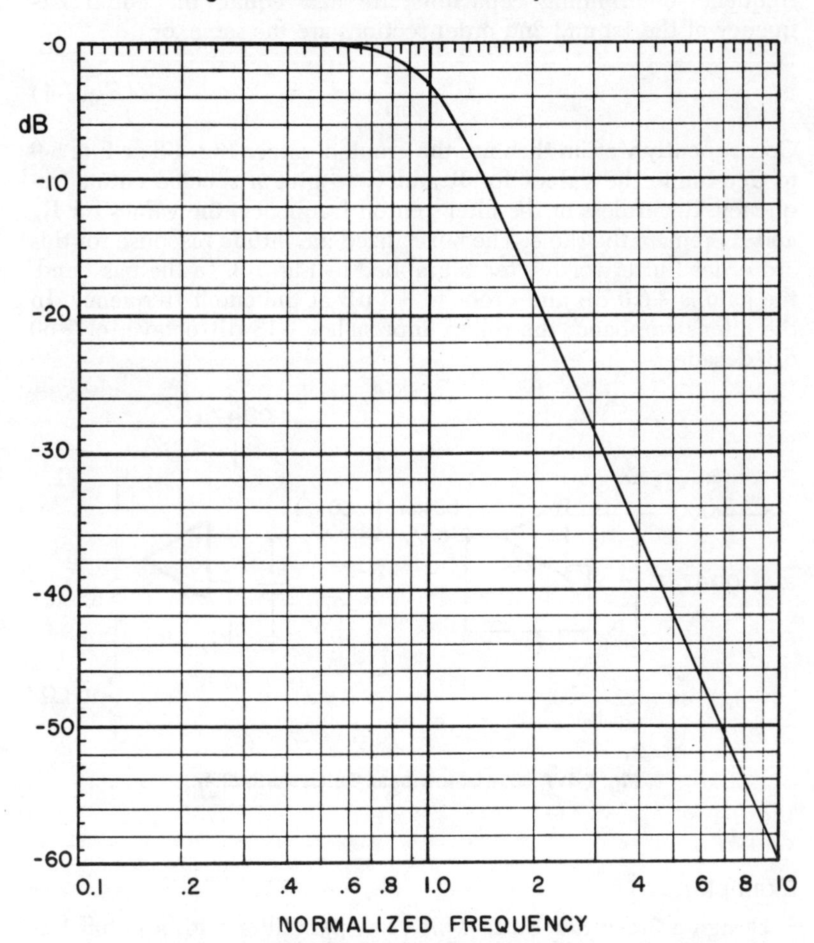

Fig. 7-3. Normalized amplitude response for filter of Fig. 7-2.

3RD ORDER FILTERS

The basic 1-kHz 3rd order low-pass filter is shown in Fig. 7-2. It consists of a unity gain 1st order section followed by an "equal-component" VCVS 2nd order section.

Since we are now concerned with the *design* of useful filter circuits, the 3rd order low-pass filter is normalized for a cutoff frequency of 1 kHz and an impedance level of 10 kΩ. The resistors R_A and R_B set the required gain of the 2nd order section at 2.0, as indicated in Table 7-2 to give a damping factor of 1.000 (Table 7-1). Since the frequency-determining resistors (R) are equal, and the frequency-determining capacitors are also equal, the cutoff frequency of the 1st and 2nd order sections are the same, or

$$f_c = \frac{1}{2\pi RC} \qquad \text{(Eq. 7-4)}$$

Consequently we can then use the graph of either Fig. 4-8 or Fig. 5-9 to determine the values for R and C to give a specific cutoff frequency. Regardless of the filter's cutoff frequency, the values for R_A and R_B remain the same. The normalized amplitude response for this 3rd order Butterworth filter is graphed in Fig. 7-3. In the passband, the gain is +6.0 dB and drops to +3 dB at the cutoff frequency. In the filter's stopband, the rolloff approaches −18 dB/octave, or −60 dB/decade.

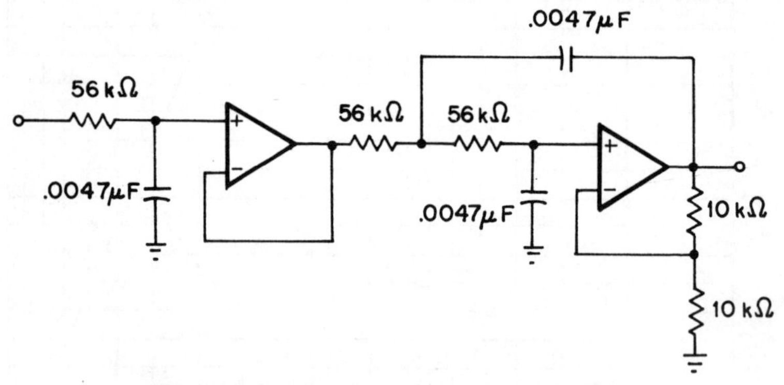

Fig. 7-4. A 600-Hz, low-pass Butterworth filter.

Example:

Design a 3rd order Butterworth low-pass filter with a cutoff frequency of 600 Hz.

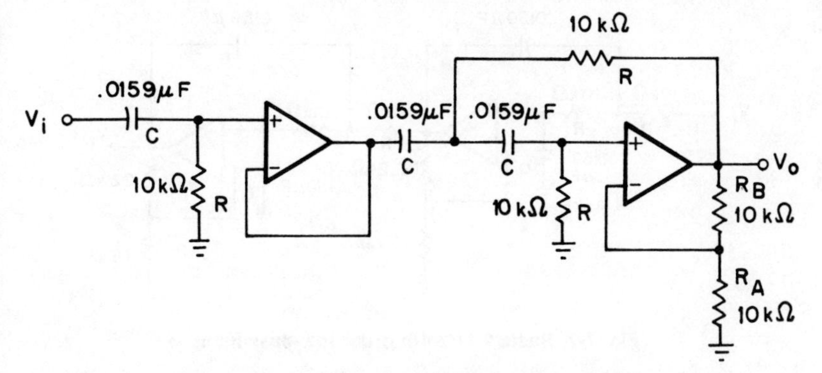

Fig. 7-5. A 3rd order high-pass Butterworth filter.

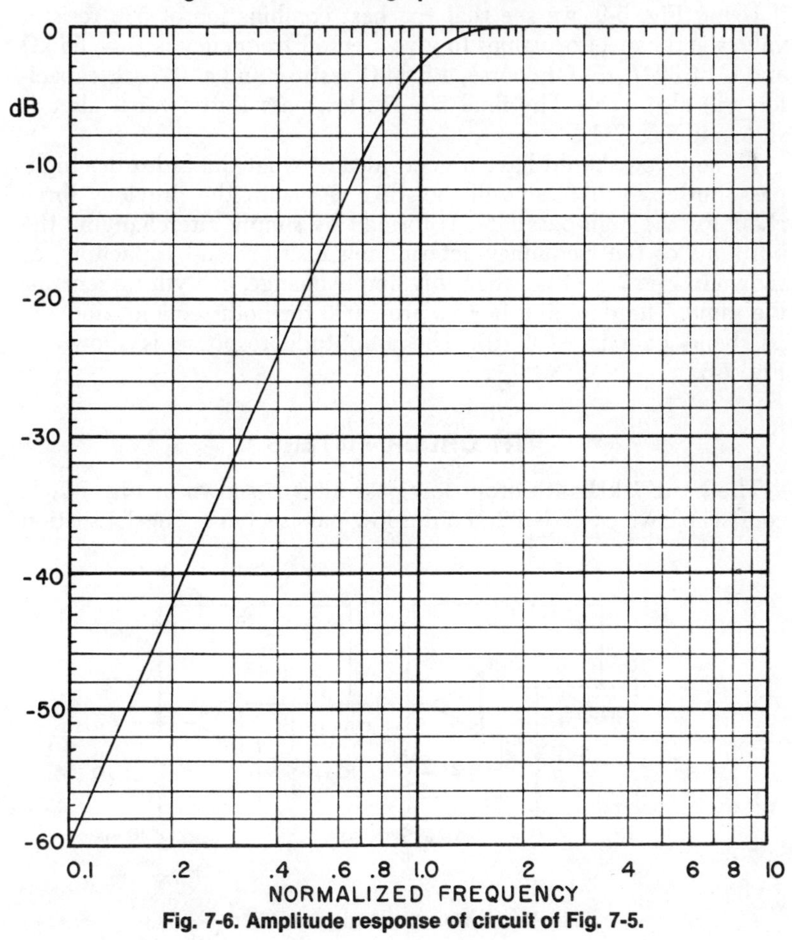

Fig. 7-6. Amplitude response of circuit of Fig. 7-5.

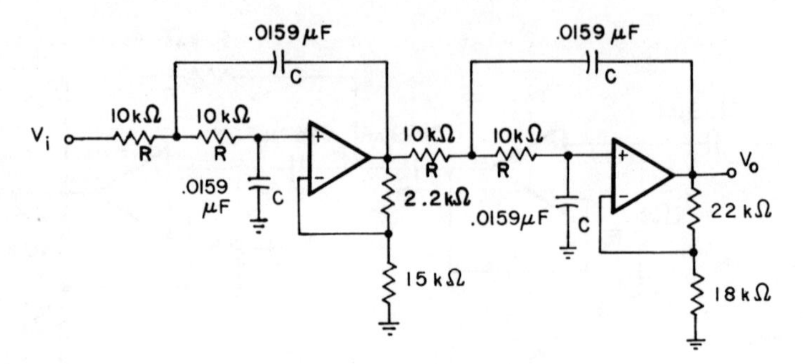

Fig. 7-7. Basic 1-kHz 4th order low-pass filter.

Using Fig. 5-9, we see that the best combination of 5% resistor values and capacitor values to give a cutoff frequency is R = 56 kΩ and C = .0047 μF. Of course, a 5.6-kΩ resistor and a .047-μF capacitor will also work. The final 600-Hz low-pass Butterworth filter is shown in Fig. 7-4 .

By now you should have a good idea of what the 3rd order high-pass Butterworth filter will look like. As with the previous three chapters, the high-pass filter is formed by simply interchanging the positions of the frequency-determining resistors and capacitors, as shown in Fig. 7-5. Other than this simple change, everything remains the same. The damping factors are 1.000 for both sections, and the passband gain is +6.0 dB. The amplitude response is shown in Fig. 7-6.

4TH ORDER FILTERS

The basic 1-kHz 4th order low-pass filter is shown in Fig. 7-7. It consists of two cascaded 2nd order low-pass sections. The 1st section

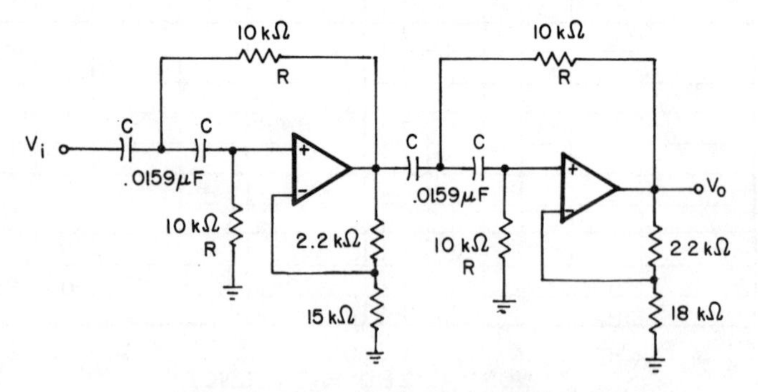

Fig. 7-8. A 4th order high-pass filter.

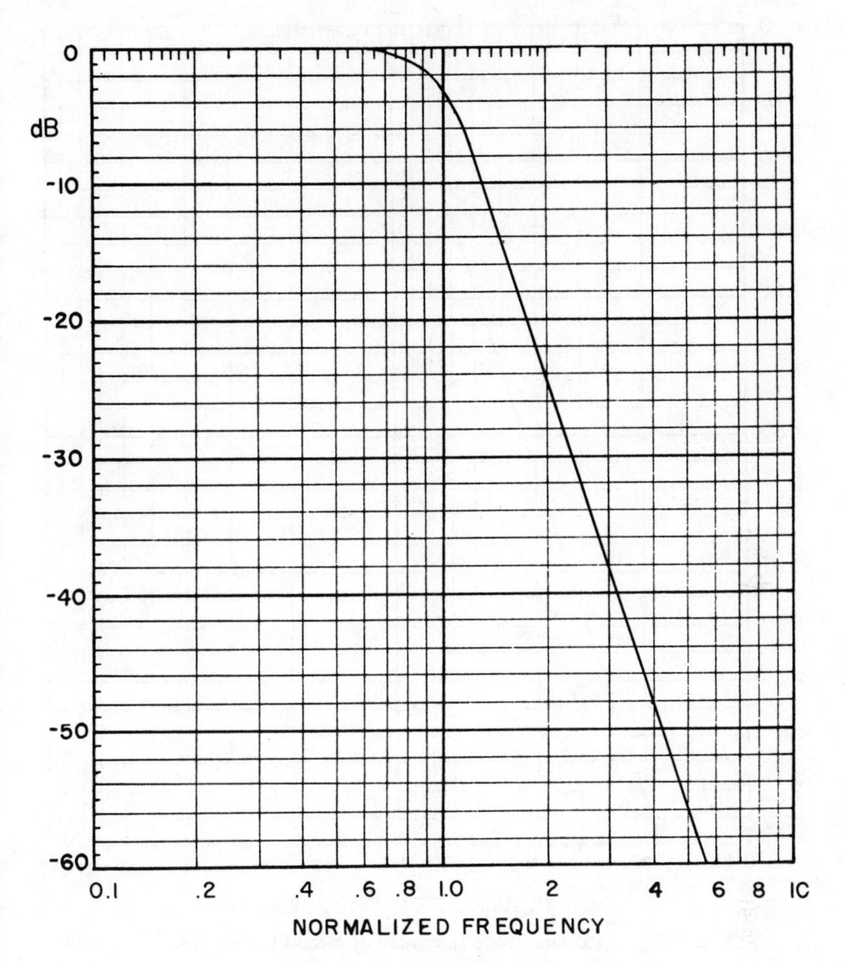

Fig. 7-9. Amplitude response of circuit in Fig. 7-7.

has the higher damping factor (1.848) and its gain is +1.2 dB. The 2nd section has a damping factor of 0.765 and a gain of +7.0 dB. As with the 3rd order filters, the frequency-determining resistors and capacitors are determined easiest from graph of Fig. 5-9. The 4th order high-pass filter is shown in Fig. 7-8.

The amplitude response for the low-pass filter is graphed in Fig. 7-9, and the high-pass response is shown in Fig. 7-10. Both filters have a passband gain of +8.2 dB, so that the cutoff frequency occurs when the amplitude response drops to +5.2 dB. For the low-pass filter, the rolloff is −24 dB/octave (−80 dB/decade), while the high-pass version has a +24 dB/octave (+80 dB/decade) rolloff.

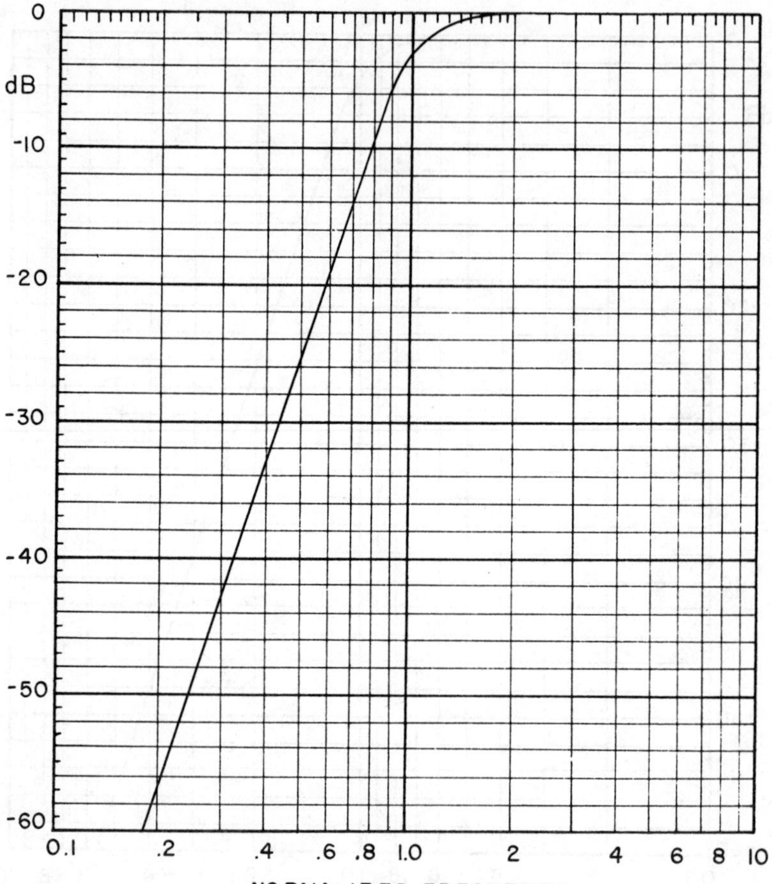

Fig. 7-10. Amplitude response of circuit in Fig. 7-8.

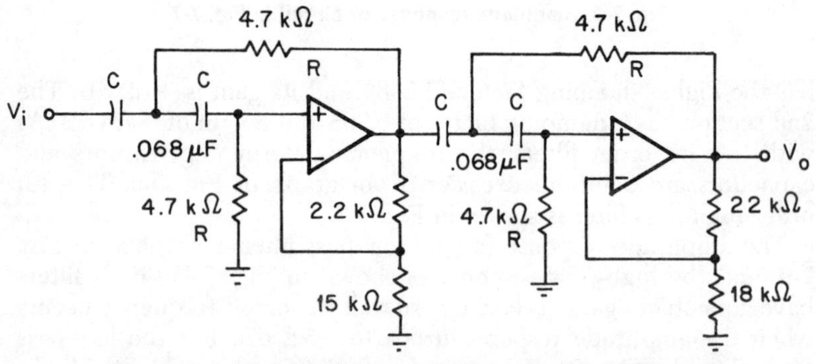

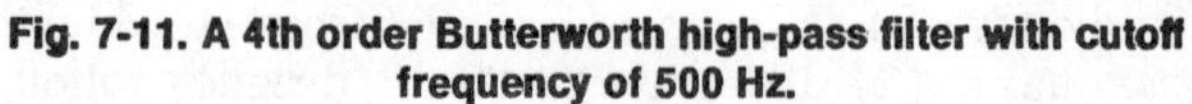
Fig. 7-11. A 4th order Butterworth high-pass filter with cutoff frequency of 500 Hz.

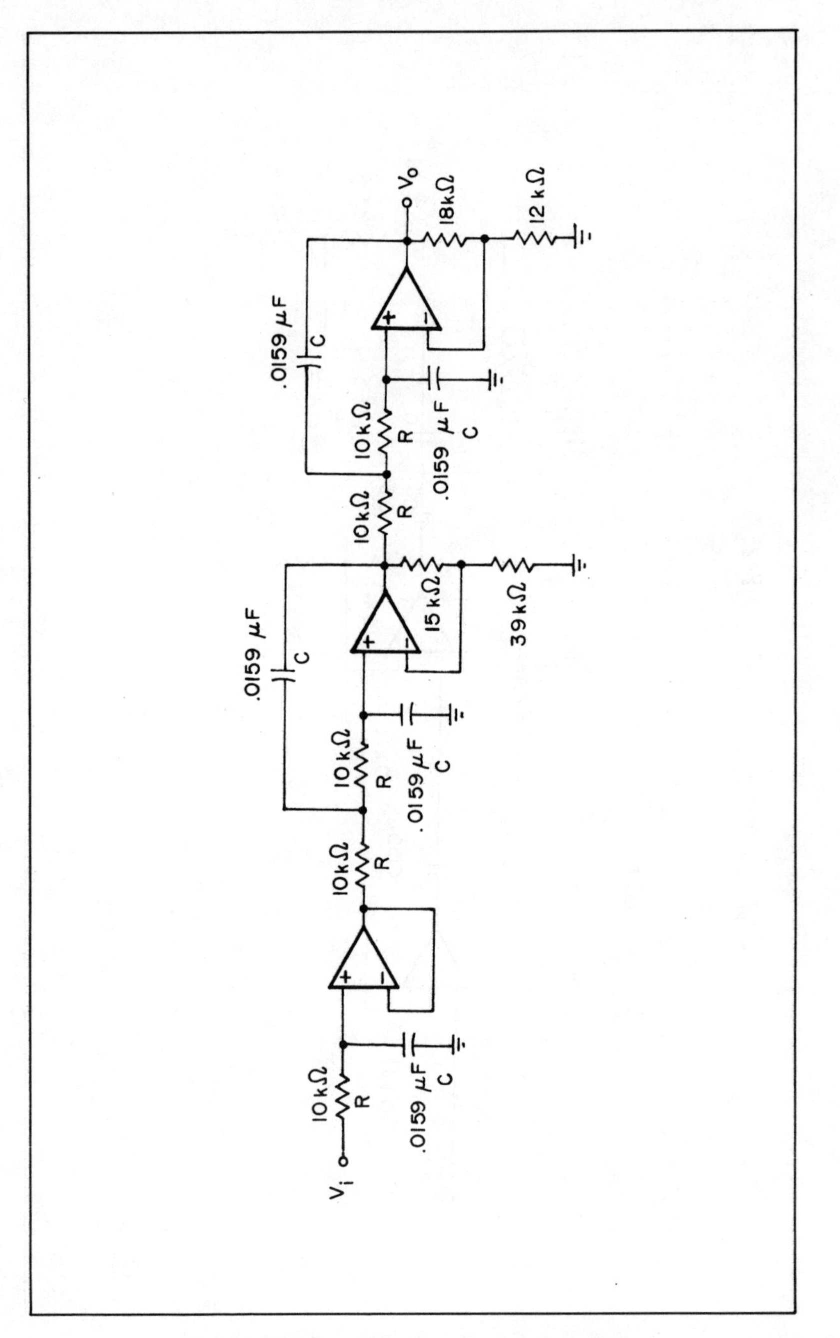

Fig. 7-12. Basic 1-kHz 5th order low-pass filter.

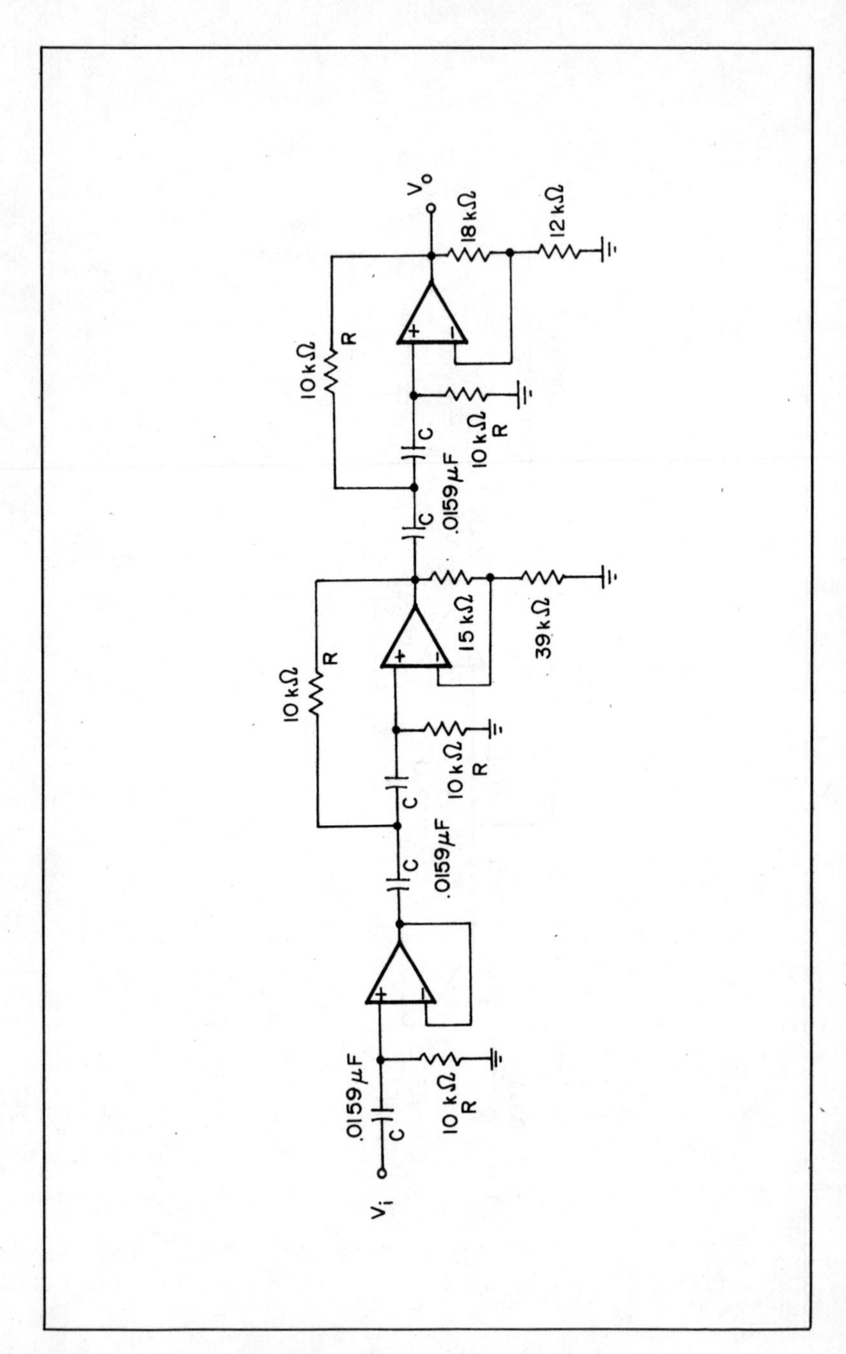

Fig. 7-13. A 5th order high-pass filter.

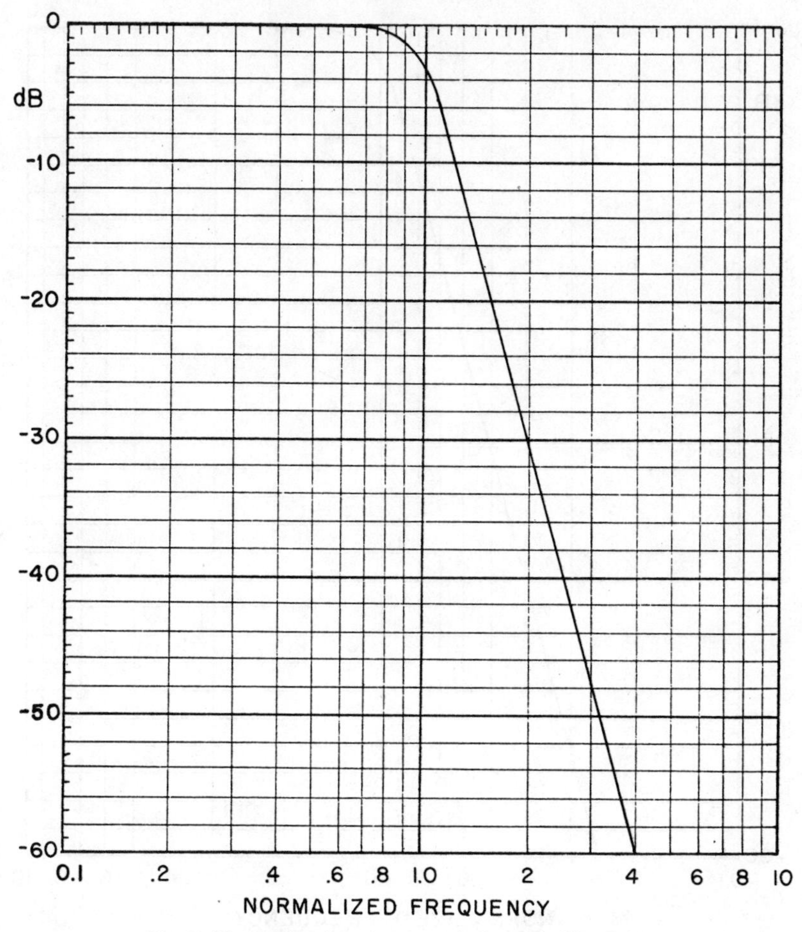

Fig. 7-14. Amplitude response of circuit in Fig. 7-12.

Example:

Design a 4th order Butterworth high-pass filter with a cutoff frequency of 500 Hz.

Using Fig. 5-9, we see that the best combination for the 500-Hz cutoff frequency is R = 47 kΩ and C = .0068 μF, or R = 4.7 kΩ and C = .068 μF, as shown in Fig. 7-11.

5TH ORDER FILTERS

The basic 1-kHz 5th order low-pass filter is shown in Fig. 7-12, and the 5th order high-pass filter is shown in Fig. 7-13. Both consist of a single 1st order section and two different 2nd order sections. The

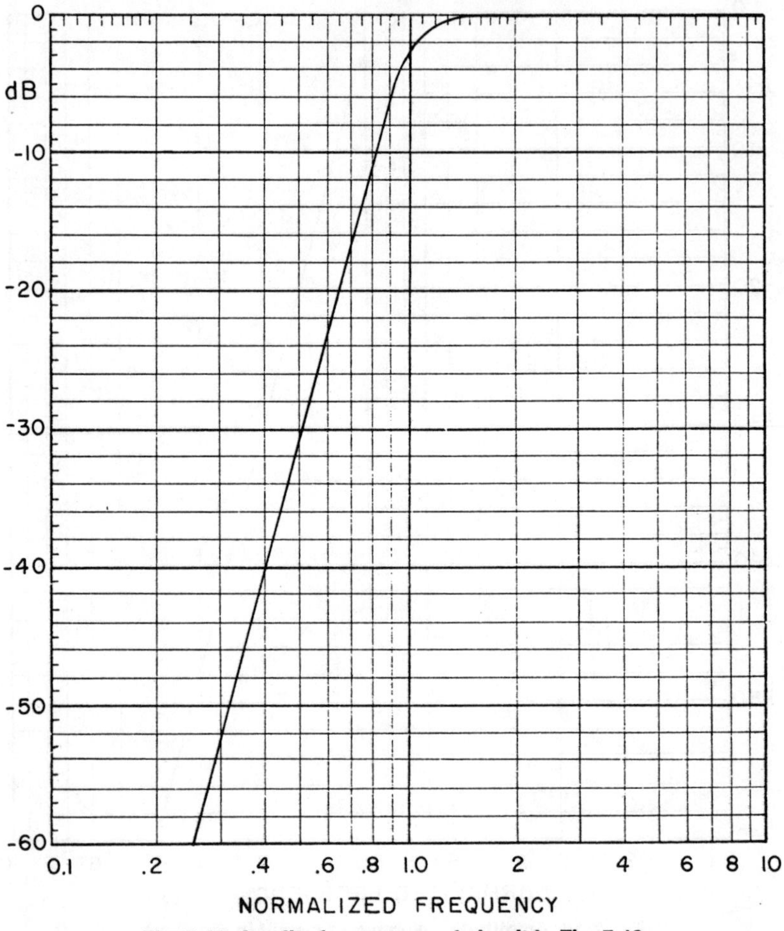

Fig. 7-15. Amplitude response of circuit in Fig. 7-13.

passband gain for both filters is fixed at +10.3 dB, so that the cutoff frequency occurs when the amplitude response drops to +7.3 dB. The 5th order low-pass response is shown in Fig. 7-14, while the high-pass response is shown in Fig. 7-15.

6TH ORDER FILTERS

The basic 1-kHz 6th order low-pass filter is shown in Fig. 7-16, and the high-pass filter is shown in Fig. 7-17. Both filters are formed by cascading three different 2nd order sections according to the damping factors given in Table 7-1. For both filters, the passband gain is fixed at +12.5 dB, and the cutoff frequency occurs at the

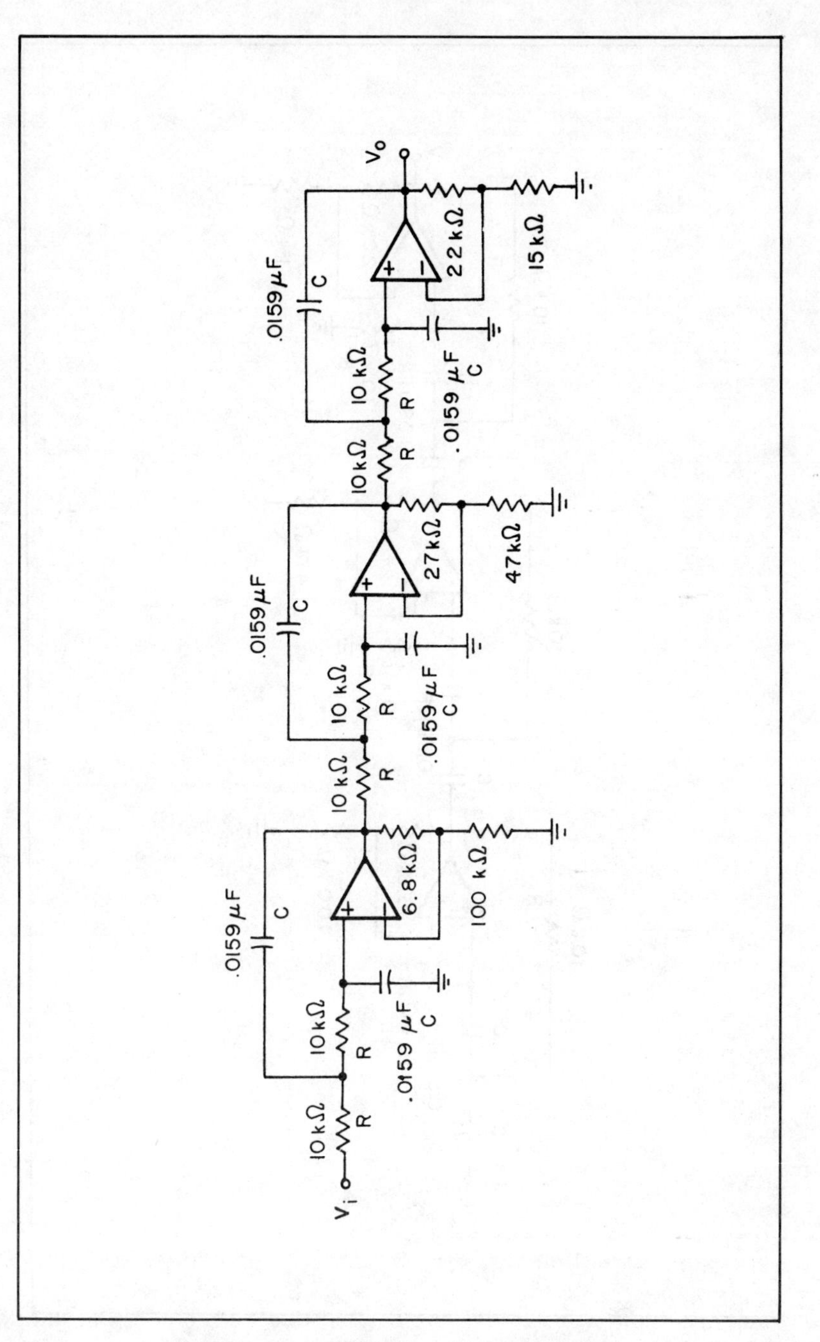

Fig. 7-16. Basic 1-kHz 6th order low-pass filter.

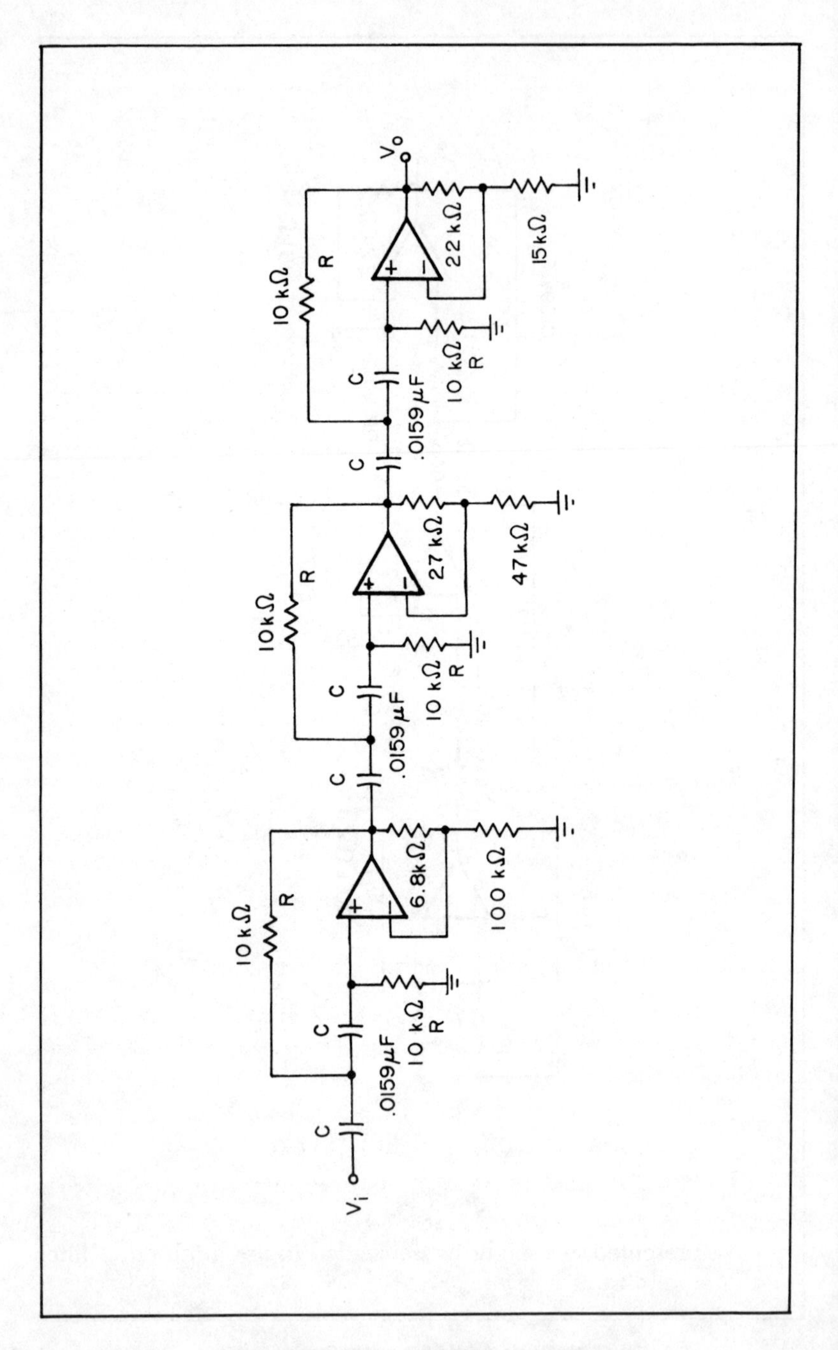

Fig. 7-17. A 6th order high-pass filter.

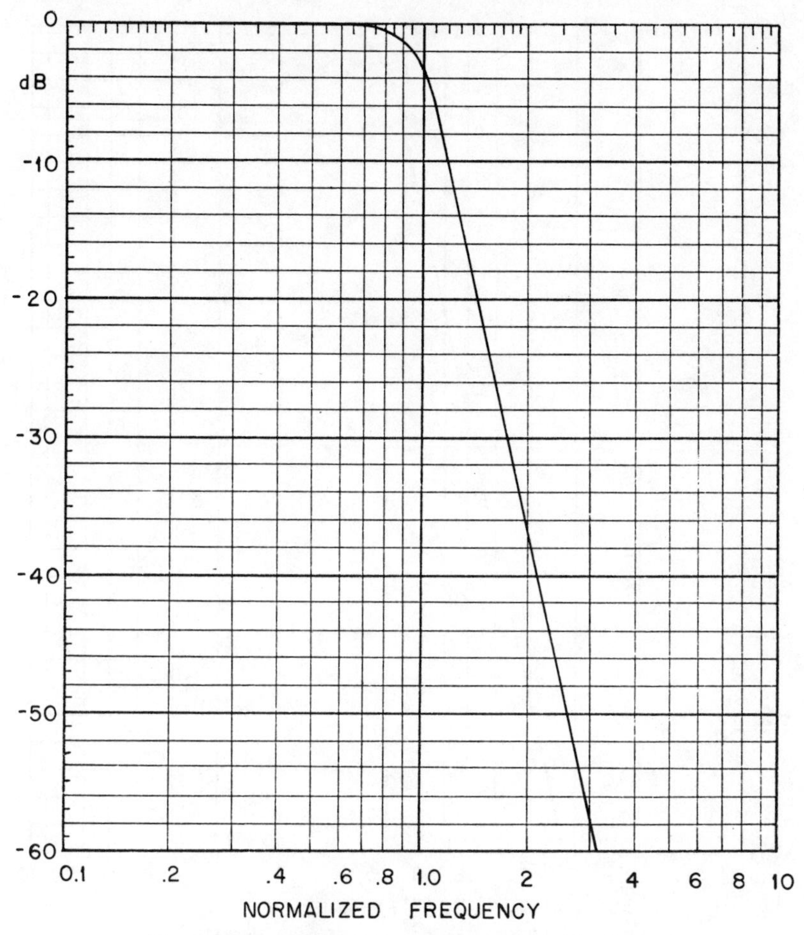

Fig. 7-18. Amplitude response of circuit in Fig. 7-16.

point when the filter's amplitude response drops to +9.5 dB. The 6th order low-pass amplitude response is graphed in Fig. 7-18, and the high-pass response is shown in Fig. 7-19.

WHICH ORDER FILTER

By now you have probably thought, "which order filter do I need?" Since the 6th order filter has the best response of all the filters we have presented, we might be persuaded to use a 6th order filter for all our applications. However, when we use a higher order filter than is really necessary, we waste a lot of space on the circuit board in addition to paying more for what is not needed.

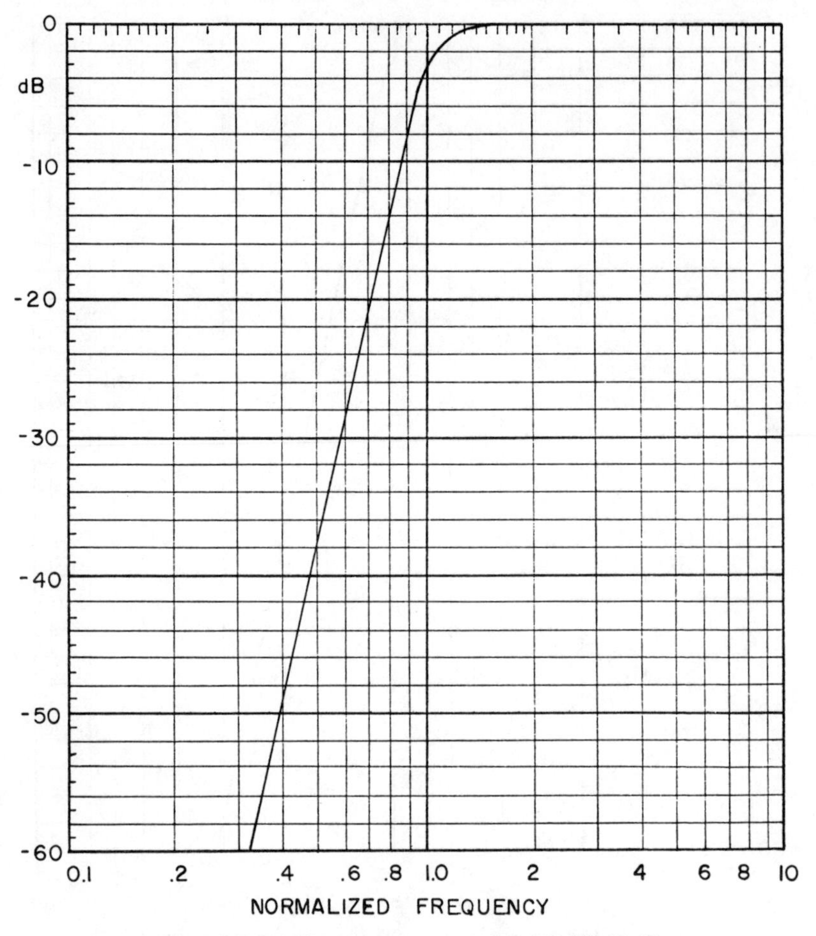

Fig. 7-19. Amplitude response of circuit in Fig. 7-17.

In general, the minimum order required is dictated by the specific requirements that are imposed. As an example, suppose we need a filter that will pass frequencies up to approximately 1 kHz while rejecting a 3-kHz signal by 25 dB. That is, for a 3-kHz signal with an amplitude of 1-volt peak-to-peak as the input, the output signal will be 56 millivolts peak-to-peak. So far, it is safe to say that a 1-kHz low-pass filter is needed, but, what order low-pass filter? In Fig. 7-20, the stopband responses of 1st through 6th order low-pass filters are compared up to two decades above a normalized cutoff frequency of 1.0. In this case, a frequency of 3 kHz corresponds to a normalized frequency of 3.0, so that −25 dB lies between the response of the 2nd and 3rd order filters. Therefore, to have at least

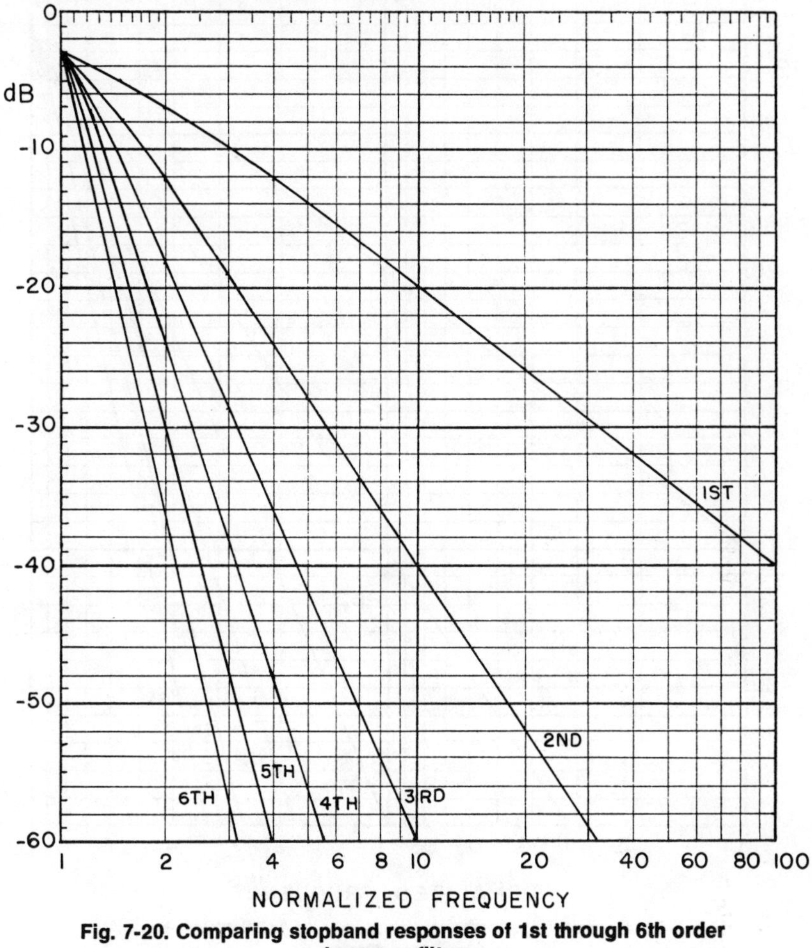

Fig. 7-20. Comparing stopband responses of 1st through 6th order low-pass filters.

25 dB of rejection at 3 kHz, we must use a 3rd order 1-kHz low-pass filter. The graph of Fig. 7-21 compares the stopband responses for high-pass filters.

AN INTRODUCTION TO THE EXPERIMENTS

The following experiments are designed to demonstrate the design and operation of higher order Butterworth active filters. Using 5% resistors and 10% capacitors, you should have little trouble in obtaining respectable results for 3rd and 4th order filters. For higher order filters, such as 5th and 6th order, we would naturally require better than 5% resistors and 10% capacitors. In an effort to keep the

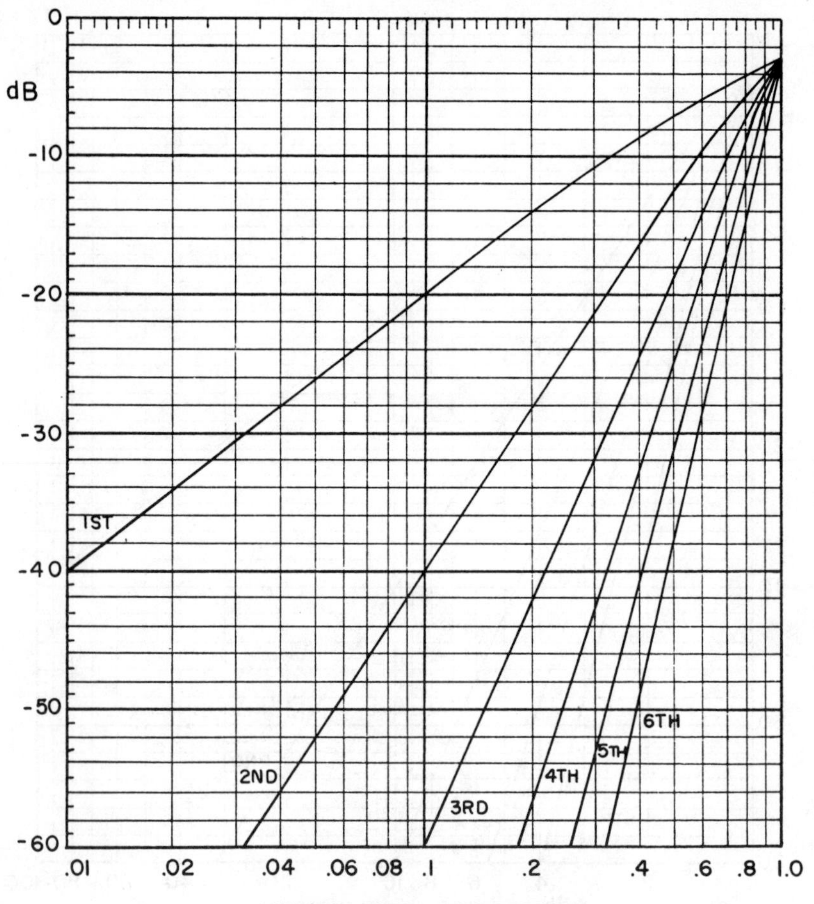

Fig. 7-21. Comparing stopband responses of 1st through 6th order high-pass filters.

expenses down, we will use 5% resistors and 10% capacitors, and with a little luck, hope to obtain adequate results.

The experiments that you will perform can be summarized as follows:

Experiment No.	*Purpose*
1	Demonstrates the operation and design of a 3rd order low-pass Butterworth filter.
2	Demonstrates the operation and design of a 4th order high-pass Butterworth filter.

3 Demonstrates the operation and design of a 5th order low-pass Butterworth filter.

EXPERIMENT NO. 1

Purpose

The purpose of this experiment is to demonstrate the operation and design of a 3rd order low-pass Butterworth active filter.

Schematic Diagram of Circuit (Fig. 7-22)

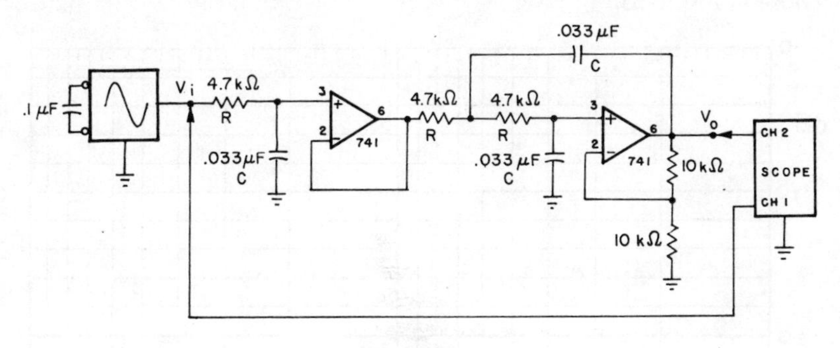

Fig. 7-22.

Design Basics

- 1st Order Section:
    - Cutoff frequency: $f_c = 1/2\pi RC$
    - Damping: $\alpha = 1.000$
    - Passband gain: unity (0 dB)
- 2nd Order Section:
    - Cutoff frequency: $f_c = 1/2\pi RC$
    - Damping: $\alpha = 1.000$
    - Passband gain: $3 - \alpha = 2.000$ (+6.02 dB)
- Amplitude response: $20 \log_{10}\left[\frac{2.00}{[1 + (f)^6]^{1/2}}\right]$

Step 1

Set your oscilloscope for the following settings:

- Channel 1: 500 mV/division
- Channel 2: 500 mV/division
- Time base: 1 ms/division
- AC coupling

Step 2

Wire the circuit shown in the schematic diagram. Apply power to the breadboard and adjust the input voltage to 1 volt peak-to-peak and the frequency at 100 Hz. Determine the voltage gain at 100 Hz. Is it what you expect it to be?

The passband voltage gain should be approximately 2.00.

Step 3

Vary the generator frequency and complete the following table and plot your results on the blank graph page provided for this purpose in Fig. 7-23.

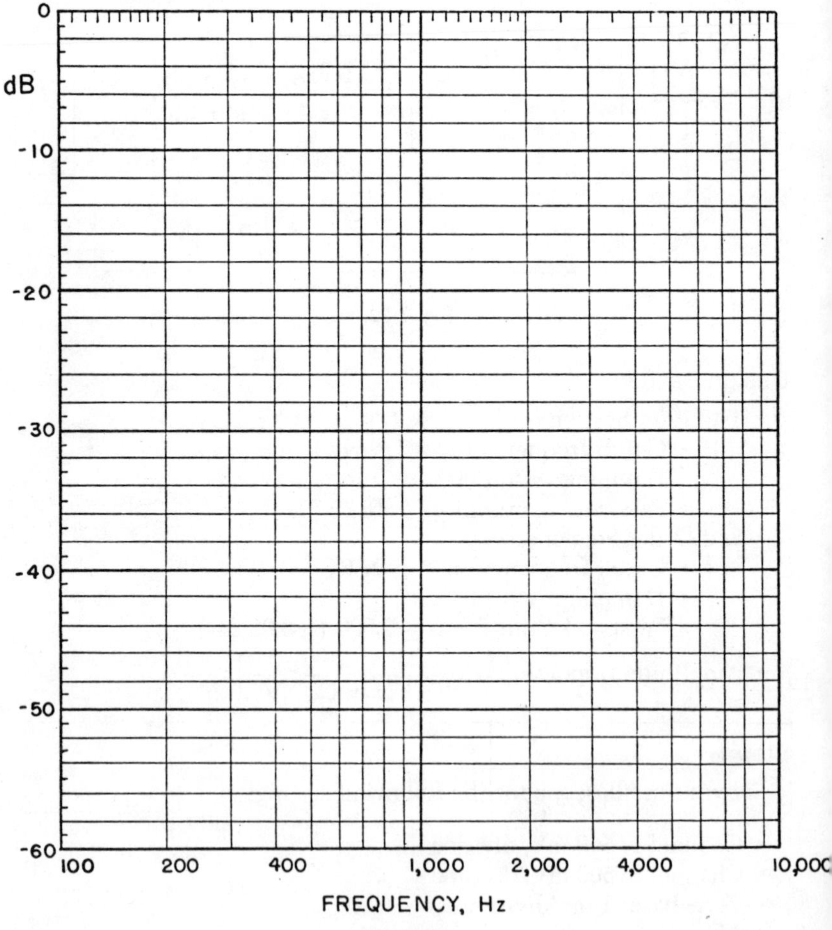

Fig. 7-23.

Frequency	V_o	V_o/V_i	Experimental dB Gain	Theoretical dB Gain*
100 Hz				+6.02
200				+6.02
400				+6.01
600				+5.85
800				+5.14
1000				+3.44
2000				−11.4
4000				−29.4
8000				−47.5

* Based on a cutoff frequency of 1027 Hz. For simplicity, we'll assume the cutoff frequency to be 1000 Hz.

From either the table or your graph, at what approximate frequency is the dB gain 3 dB less than the passband gain? Is it what you expected?

As a check on your results, you may want to compare them with mine, listed below.

Frequency	V_o	V_o/V_i	Experimental dB Gain	Theoretical dB Gain
100 Hz	1.95 V	1.95	+5.80	+6.02
200	1.95	1.95	+5.80	+6.02
400	1.95	1.95	+5.80	+6.01
600	1.85	1.85	+5.34	+5.85
800	1.66	1.66	+4.40	+5.14
1000	1.36	1.36	+2.67	+3.44
2000	0.250	0.25	−12.0	−11.4
4000	0.031	0.031	−30.0	−29.4
8000	0.004	0.004	−48.0	−47.5

Step 4

From your results, what is the rolloff in the stopband?

A 3rd order filter has a stopband rolloff of −18 dB/octave, or −60 dB/decade. In this experiment, this rolloff is best seen by taking the difference of the values at 4000 Hz and 8000 Hz, or 1 octave.

Step 5

As an optional exercise, select some particular cutoff frequency, and with the aid of Fig. 5-9, determine the proper values for the frequency-determining resistors and capacitors. Then repeat this experiment.

EXPERIMENT NO. 2

Purpose

The purpose of this experiment is to demonstrate the operation and design of a 4th order high-pass Butterworth active filter.

Schematic Diagram of Circuit (Fig. 7-24)

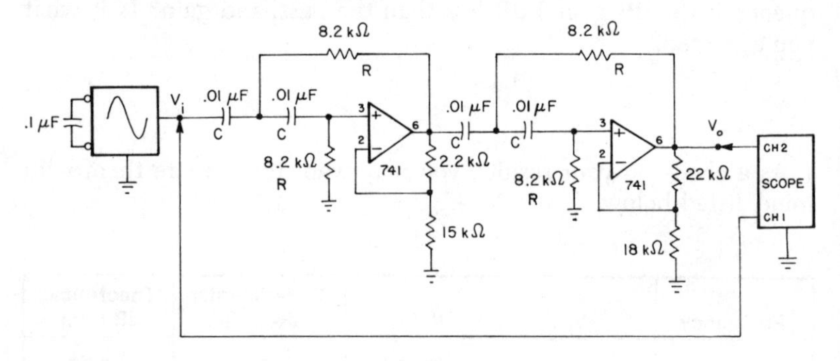

Fig. 7-24.

Design Basics

- First 2nd Order Section:
  - Cutoff frequency: $f_c = 1/2\pi RC$
  - Damping: $\alpha = 1.848$
  - Passband gain: 1.152 (+1.23 dB)
- Second 2nd Order Section:
  - Cutoff frequency: $f_c = 1/2\pi RC$
  - Damping: $\alpha = 0.765$
  - Passband gain: 2.235 (+6.99 dB)
- Amplitude response: $20 \log_{10} \left[\frac{2.57}{[1 + (f)^8]^{1/2}}\right]$

Step 1

Set your oscilloscope for the following settings:

- Channel 1: 500 mV/division
- Channel 2: 10 mV/division
- Time base: 1 ms/division
- AC coupling

Step 2

Wire the circuit in the schematic diagram. Apply power to the breadboard and adjust the input voltage to 2 volts peak-to-peak and the frequency to 500 Hz (5 full cycles).

Step 3

Vary the generator frequency to complete the following table and plot your results on the blank graph provided for this purpose in Fig. 7-25.

Frequency	V_o	V_o/V_i	Experimental dB Gain	Theoretical dB Gain*
500 Hz				−38.9
600				−32.6
700				−27.3
800				−22.6
900				−18.5
1000				−14.9
2000				+5.67
3000				+8.07
4000				+8.19
5000				+8.20
6000				+8.20

* Based on a cutoff frequency of 1941 Hz. For simplicity, we'll assume the cutoff frequency to be 2 kHz.

Step 4

Determine the filter's stopband rolloff from your results by subtracting the dB value measured at 500 Hz from the measured value at 1000 Hz (1 octave). Is it what you would expect from a 4th order high-pass filter with a Butterworth response?

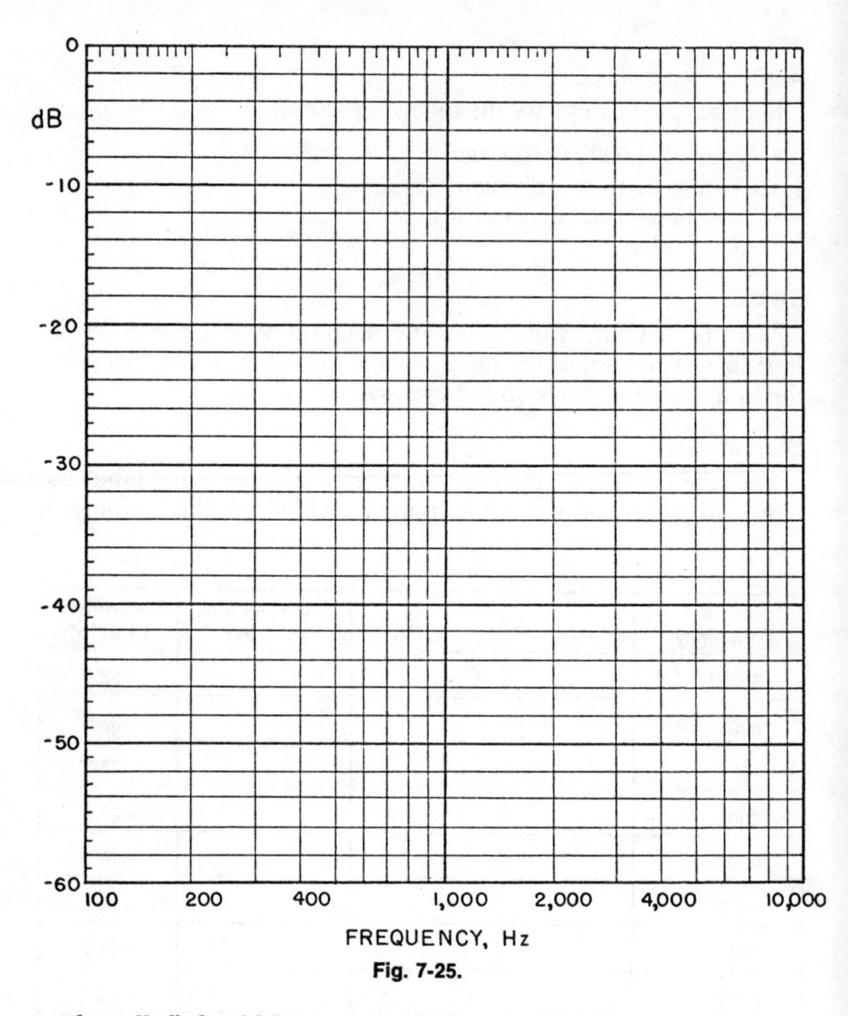

Fig. 7-25.

The rolloff should be approximately +24 dB/octave.

EXPERIMENT NO. 3

Purpose

The purpose of this experiment is to demonstrate the operation and design of a 5th order low-pass Butterworth active filter.

Schematic Diagram of Circuit (Fig. 7-26)

Design Basics

- 1st Order Section:
    - Cutoff frequency: $f_c = 1/2\pi RC$

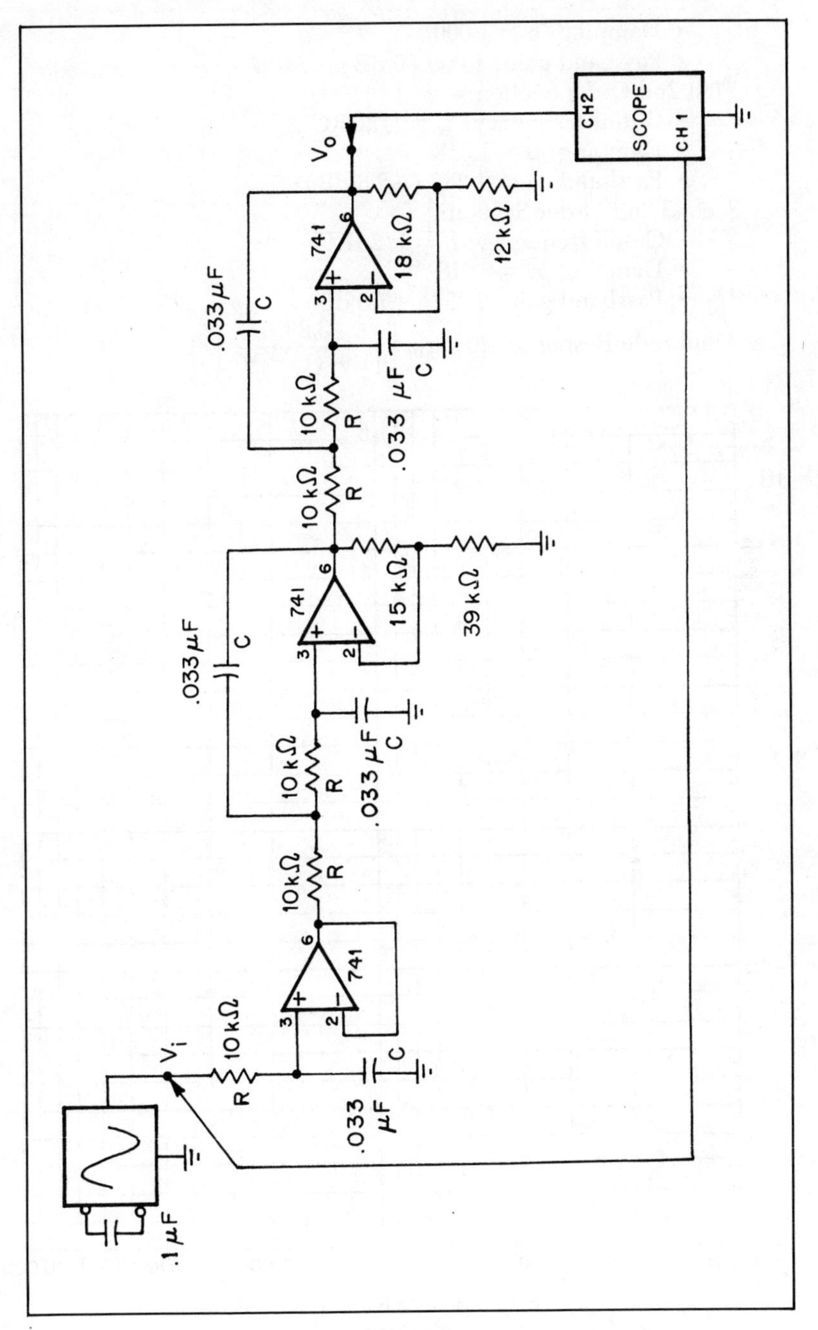
.1 µF
V_i
R
10 kΩ
.033 µF
C
3
2
6
741
10 kΩ
R
10 kΩ
R
.033 µF
C
.033 µF
C
741
15 kΩ
39 kΩ
10 kΩ
R
10 kΩ
R
.033 µF
C
.033 µF
C
741
18 kΩ
12 kΩ
V_o
CH2
SCOPE
CH1

Fig. 7-26.

- Damping: $\alpha = 1.000$
- Passband gain: 1.000 (0 dB)

- First 2nd Order Section:
  - Cutoff frequency: $f_c = 1/2\pi RC$
  - Damping: $\alpha = 1.618$
  - Passband gain: 1.382 (+2.8 dB)
- Second 2nd Order Section:
  - Cutoff frequency: $f_c = 1/2\pi RC$
  - Damping: $\alpha = 0.618$
  - Passband gain: 2.382 (+7.5 dB)
- Amplitude Response: $20 \log_{10} \left[\frac{3.29}{[1 + (f)^{10}]^{1/2}} \right]$

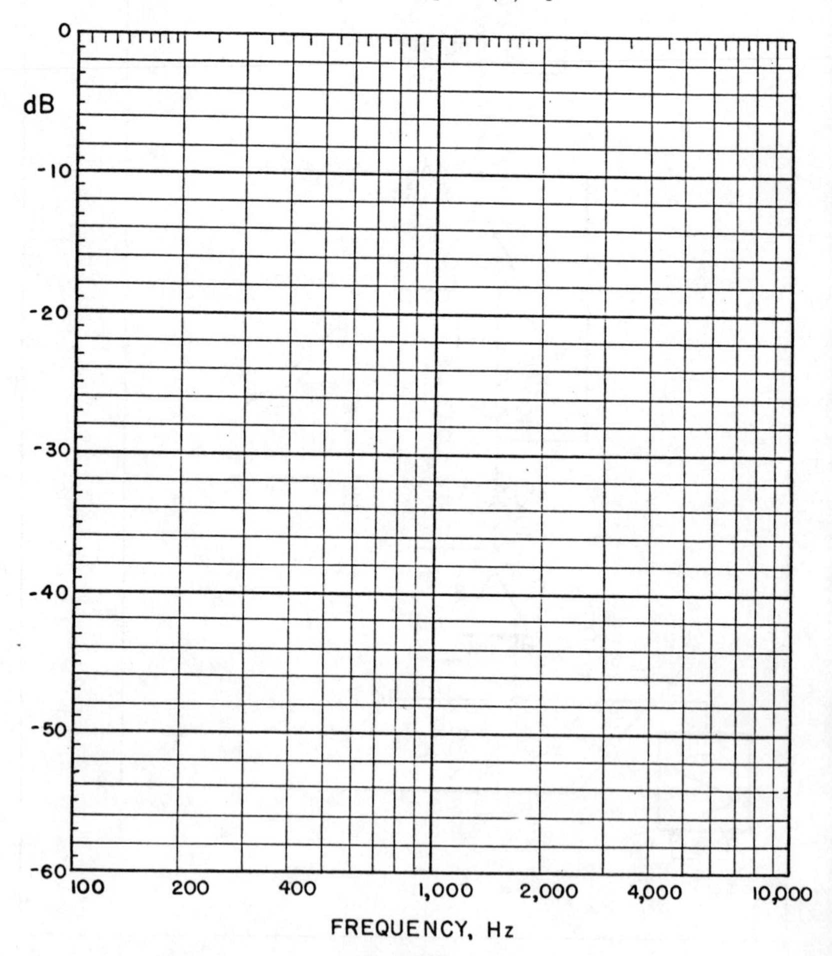

Fig. 7-27.

Step 1

Set your oscilloscope for the following settings:

- Channel 1: 500 mV/division
- Channel 2: 1 V/ division
- Time base: 1 ms/division
- AC coupling

Step 2

Wire the circuit in the schematic diagram. Apply power to the breadboard and adjust the input voltage at 2 volts peak-to-peak, and the generator frequency at 100 Hz (1 full cycle/10 horizontal divisions).

Step 3

Starting at 100 Hz, vary the generator frequency to complete the following table, and plot your results on the blank graph provided for this purpose in Fig. 7-27.

Frequency	V_o	V_o/V_i	Experimental dB Gain	Theoretical dB Gain
100 Hz				+10.34
200				+10.34
300				+10.32
400				+9.90
500				+7.33
600				+1.78
700				−4.42
800				−10.11
900				−15.20
1000				−19.76
2000*				−49.86

* You may have difficulty in measuring the peak-to-peak output voltage at this frequency. However, make the best measurement you can.

Step 4

Determine the filter's stopband rolloff from your results by subtracting the dB value measured at 2000 Hz from the measured value at 1000 Hz (1 octave). Is it what you would expect from a 5th order low-pass Butterworth filter?

The rolloff should be 5 times −6 dB/octave, or −30 dB/octave, which is the same as −100 dB/decade.

If we were able to accurately measure the response at 5 kHz, we would expect an amplitude response of approximately −89.7 dB.

CHAPTER 8

Active Bandpass and Notch Filters

INTRODUCTION

In this chapter, several basic circuits covering the design and performance of bandpass and notch filters are presented. Although there are a number of possible circuits that will realize the bandpass and notch filter responses, the networks presented in this chapter probably represent the best compromise between design and performance.

OBJECTIVES

At the completion of this chapter, you will be able to do the following:

- Design multiple-feedback bandpass filters, using a single operational amplifier
- Properly cascade identical bandpass sections for greater stopband rolloff
- Design wideband filters
- Design notch filters

THE MULTIPLE-FEEDBACK BANDPASS FILTER

As shown in Fig. 8-1, the basic *multiple-feedback bandpass filter* is useful for Q's up to approximately 15-20 with "moderate" gains. For this circuit, the center frequency is determined from the relation

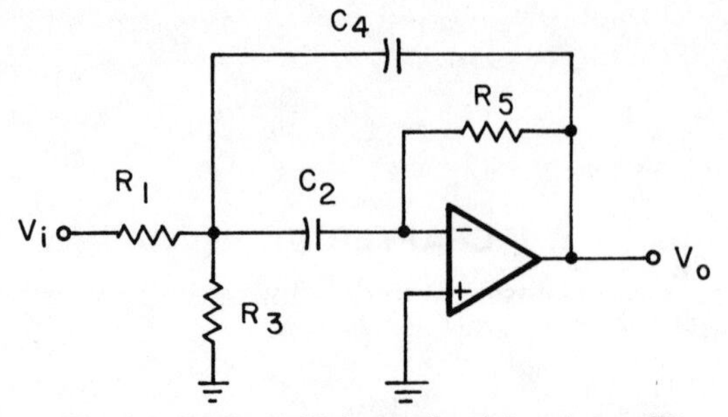

Fig. 8-1. Basic multiple-feedback bandpass filter.

$$\omega_o = \left[\frac{1}{R_5 C_2 C_4}\left(\frac{1}{R_1} + \frac{1}{R_3}\right)\right]^{1/2} \qquad \text{(Eq. 8-1)}$$

As a function of the filter's passband gain G, and Q, the five components are found from the equations

$$R_1 = \frac{Q}{GC_4\omega_o} \qquad \text{(Eq. 8-2)}$$

$$R_3 = \frac{Q}{(2Q^2 - G)C_4\omega_o} \qquad \text{(Eq. 8-3)}$$

and

$$R_5 = \frac{2Q}{C_4\omega_o} \qquad \text{(Eq. 8-4)}$$

The filter's passband gain, i.e. the gain at the filter's center frequency, is

$$G = \frac{R_5}{R_1\left[1 + \frac{C_4}{C_2}\right]} \qquad \text{(Eq. 8-5)}$$

The selection of the five component values is eased by making C_2 and C_4 *equal*, so that

$$R_1 = \frac{Q}{GC\omega_o} \qquad \text{(Eq. 8-6)}$$

$$R_3 = \frac{Q}{(2Q^2 - G)C\omega_o} \qquad \text{(Eq. 8-7)}$$

$$R_5 = \frac{2Q}{C\omega_o} \qquad \text{(Eq. 8-8)}$$

where

$$G = \frac{R_5}{2R_1} \qquad \text{(Eq. 8-9)}$$

One nice feature of this circuit is that the center frequency can be shifted to a new frequency ω_o' while the passband gain and bandwidth remain constant, simply by changing resistor R_3 to a new value R_3', so that

$$R_3' = R_3\left[\frac{\omega_o}{\omega_o'}\right]^2 \qquad \text{(Eq. 8-10)}$$

On the other hand, because of the denominator of Equation 8-7, we are restricted to the condition

$$Q > \left[\frac{G}{2}\right]^{1/2} \qquad \text{(Eq. 8-11)}$$

Normally, we select a convenient value for C_2 and C_4, and then calculate the values for the three resistors based on the required values for Q, G, and ω_o.

Example:

Design a 750-Hz bandpass filter, using the circuit of Fig. 8-1, with $G = 1.32$ (+2.4 dB), and $Q = 4.2$. Then change the cutoff frequency to 600 Hz, holding G and the bandwidth (180 Hz) constant.

First, we pick a standard value for C_2 and C_4, both equal, such as .01 μF. Then the resistors are systematically determined from Equations 8-6 through 8-9, so that

$$R_1 = \frac{Q}{GC\omega_o} = \frac{4.2}{(2\pi)(750\ \text{Hz})(1.32)(.01\ \mu\text{F})}$$
$$= 67.6\ \text{k}\Omega$$
$$R_5 = 2R_1G = (2)(67.6\ \text{k}\Omega)(1.32)$$
$$= 178\ \text{k}\Omega$$

and

$$R_3 = \frac{Q}{(2Q^2 - G)C\omega_o} = \frac{4.2}{[(2)(4.2)^2 - 1.32](.01\ \mu\text{F})(2\pi)(750)}$$
$$= 2.6\ \text{k}\Omega$$

Using 5% resistors, the completed circuit is shown in Fig. 8-2.

To change the center frequency to 600 Hz, a new value for R_3 is calculated from Equation 8-10, so that

$$R_3' = (2.7\ \text{k}\Omega)\left[\frac{750}{600}\right]^2$$
$$= 4.2\ \text{k}\Omega$$

Using a 4.31 kΩ 5% resistor, the 600-Hz circuit is shown in Fig. 8-3.

The basic circuit of Fig. 8-1 has one major drawback: *high gain must be accompanied by high Q*, as stated by the inequality of Equa-

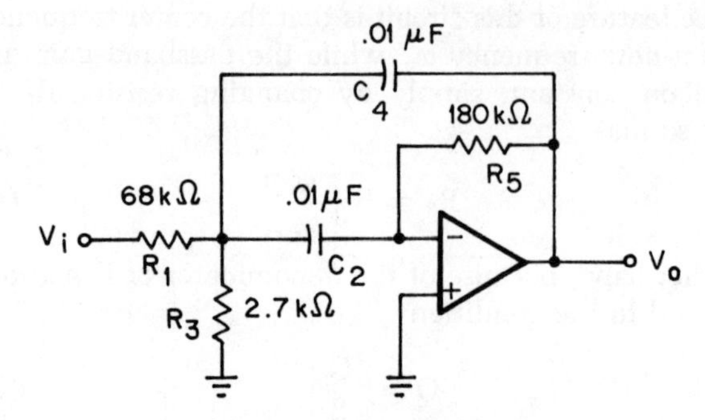

Fig. 8-2. A 750-Hz bandpass filter.

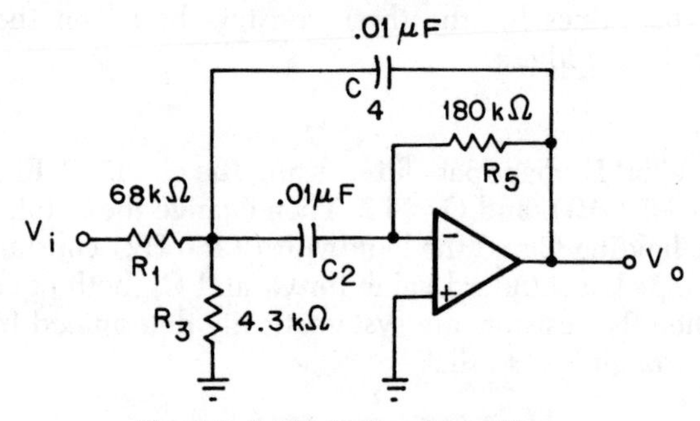

Fig. 8-3. A 600-Hz bandpass filter.

tion 8-11. Suppose this inequality does not hold? What do we do, since it is impossible for R_3 to be a *negative number?*

One approach is to change the basic circuit slightly, by omitting R_3, as shown in Fig. 8-4. Since the elimination of R_3 is the same as letting R_3 *represent an open-circuit, or infinite resistance,* equation 8-1 then simplifies to

$$\omega_o = \frac{1}{(R_1R_5C_2C_4)^{1/2}} \qquad \text{(Eq. 8-12)}$$

As before, the determination of the component values is greatly simplified by letting C_2 and C_4 be equal, so that

$$R_1 = \frac{Q}{GC\omega_o} \qquad \text{(Eq. 8-13)}$$

$$R_5 = \frac{G}{QC\omega_o} \qquad \text{(Eq. 8-14)}$$

and

$$G = Q(R_5/R_1)^{1/2} \qquad \text{(Eq. 8-15)}$$

Consequently we no longer have the restriction between gain and Q!

Sometimes there is a combination of the filter's gain and Q which permits the use of both circuits of Fig. 8-1 and Fig. 8-4 with equal resultant responses. This is when

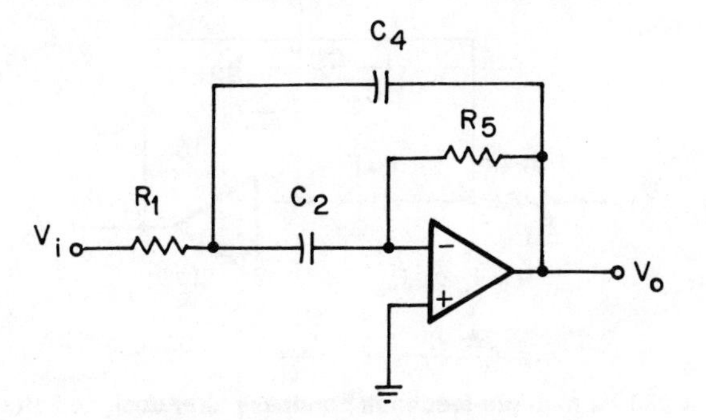

Fig. 8-4. Circuit of Fig. 8-1 without R3.

$$\frac{G}{2} < Q^2 < G \qquad \text{(Eq. 8-16)}$$

The main advantage of using Fig. 8-4 is that there is one less component. To illustrate this situation, consider the following example.

Example:

Design a 200-Hz multiple-feedback bandpass filter with a center frequency voltage gain of 30, and $Q = 5$.

From Equation 8-16, we find that it is possible to use either the circuit of Fig. 8-1, or Fig. 8-4. Using the circuit of Fig. 8-1 first, we choose a value for C_2 and C_4, for example .1 μF, so that

$$R_5 = 2Q/C\omega_o \quad \text{(Eq. 8-8)}$$

$$= \frac{(2)(5)}{(0.1\ \mu F)(2\pi)(200\ Hz)}$$

$$= 79.6\ k\Omega$$

$$R_1 = R_5/2G \quad \text{(Eq. 8-9)}$$

$$= \frac{79.6\ k\Omega}{(2)(30)}$$

$$= 1.33\ k\Omega$$

$$R_3 = Q/(2Q^2 - G)C\omega_o \quad \text{(Eq. 8-7)}$$

$$= \frac{5}{[(2)(25) - 30]\ (0.1\ \mu F)(2\pi)(200\ Hz)}$$

$$= 2.0\ k\Omega$$

Using 5% resistors, the completed design is shown in Fig. 8-5.

Now using the circuit of Fig. 8-4, we again choose a value for C_2 and C_4, for example .1 μF, so that R_1 and R_5 are found,

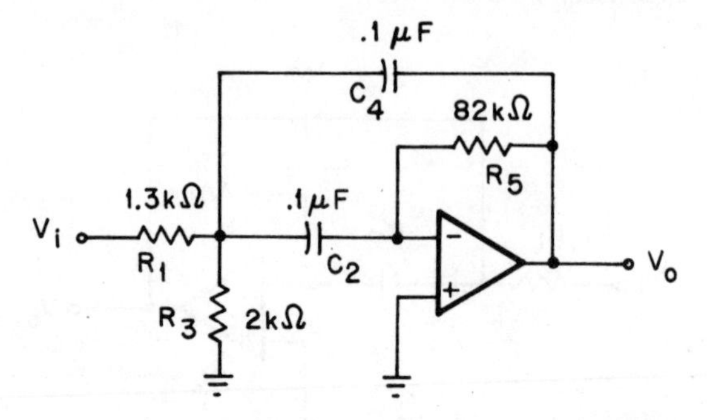

Fig. 8-5. A 200-Hz multiple-feedback bandpass filter designed after Fig. 8-1.

$$R_1 = 1.33\ \text{k}\Omega \quad (\text{Eq. 8-13})$$

$$R_5 = 47.8\ \text{k}\Omega \quad (\text{Eq. 8-14})$$

Again using 5% resistors, the completed design is shown in Fig. 8-6. It should be emphasized that *the responses of the circuits of Figs. 8-5 and 8-6 will be the same; both designs will have a center frequency of 200 Hz, a passband gain of 30 and a Q of 5.*

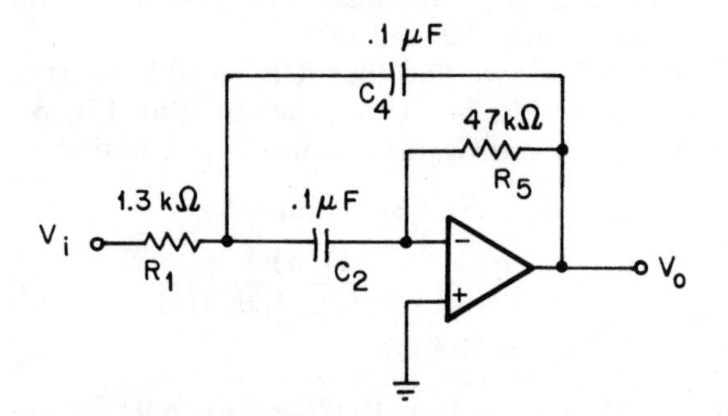

Fig. 8-6. A 200-Hz multiple-feedback bandpass filter designed after Fig. 8-4.

Now that we know how to design at least one type of bandpass filter, what will its amplitude response look like? Without any derivations, the amplitude response of a bandpass filter using a single op amp is given by

$$\text{dB} = 20 \log_{10}\text{G} - 10 \log_{10}\left[1 + \text{Q}^2\left(\frac{\omega^2 - 1}{\omega}\right)^2\right] \quad (\text{Eq. 8-17})$$

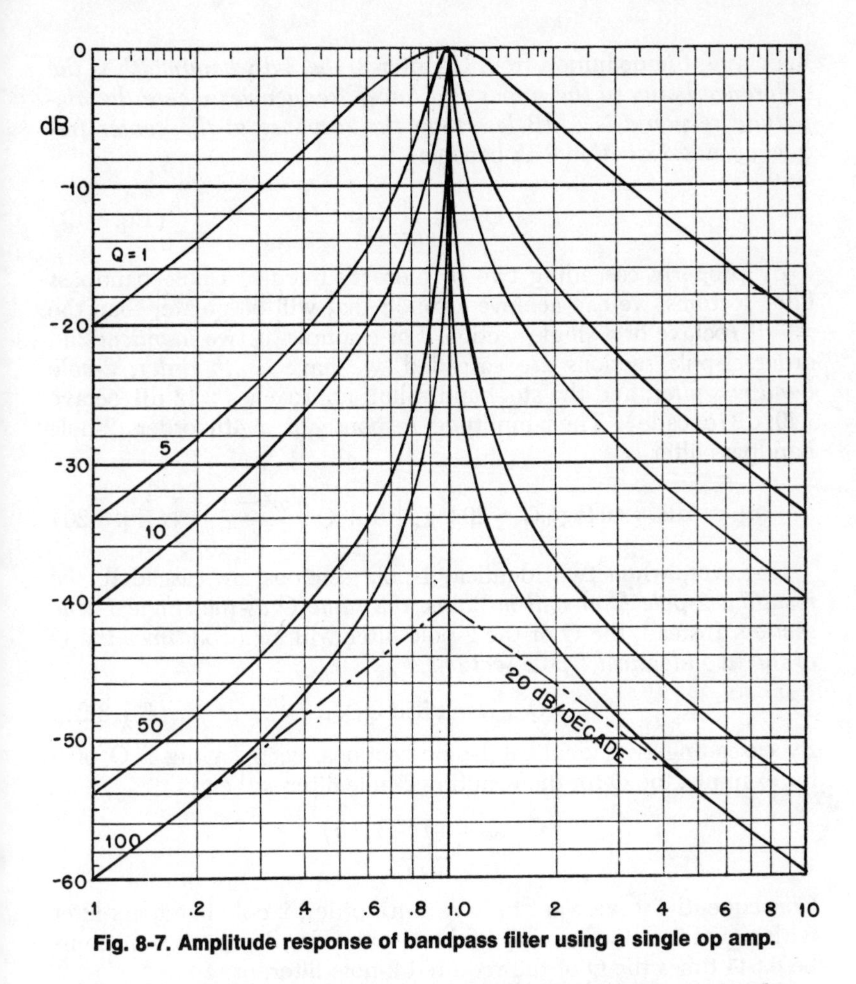

Fig. 8-7. Amplitude response of bandpass filter using a single op amp.

which is graphed in Fig. 8-7. The response is a maximum of 0 dB at the center frequency (normalized at 1.0) and then drops off on both sides. How fast the response drops is dependent on Q. However, all the curves eventually start to straighten out so that all the curves appear to be parallel. At the extremes, *the rolloff of all the curves approaches ±6 dB/octave, or ±20 dB/decade, regardless of the filter's Q*. When the response of a bandpass filter approaches a rolloff of ±6 dB/octave, it is termed a *2nd order, 1-pole bandpass filter*.

As explained earlier in Chapter 3, the *sharpness* of the bandpass filter's response in the vicinity of its center frequency depends on its Q, which in turn depends on its 3-dB *bandwidth*,

$$Q = \frac{\omega_o}{\text{3-dB bandwidth}} \qquad \text{(Eq. 8-18)}$$

Repeating the definition from Chapter 3, *the 3-dB bandwidth is the difference between the upper and lower frequencies where the amplitude response is 3 dB less than the response at the center frequency,* and Equation 8-18 becomes

$$Q = \frac{\omega_o}{\omega_H - \omega_L} \qquad \text{(Eq. 8-19)}$$

By properly cascading two or more 2nd order, 1-pole bandpass filter sections, we can achieve a rolloff that will be greater than the ±6 dB/octave of a single section. For example, if two *identical* 2nd order, 1-pole sections are cascaded, we have a *4th order, 2-pole bandpass filter,* and the stopband rolloff approaches ±12 dB/octave (40 dB/decade). The amplitude response of a 4th order, 2-pole bandpass filter is

$$dB = 40 \log_{10}G - 20 \log_{10}\left[1 + Q^2\left(\frac{\omega^2 - 1}{\omega}\right)^2\right] \qquad \text{(Eq. 8-20)}$$

However, when two identical 1-pole sections are cascaded, *the resultant 2-pole filter will not have the same Q as the single 1-pole section.* Instead, the Q of the 2-pole filter will be *1.553 times the Q of the two identical 1-pole sections,*

$$Q_{2\text{-pole}} = 1.553\, Q_{1\text{-pole}} \qquad \text{(Eq. 8-21)}$$

By cascading two identical 1-pole sections, each having a Q of 5 for example, the Q for the resulting 2-pole filter is then

$$Q_{2\text{-pole}} = (1.553)(5)$$
$$= 7.77$$

Consequently, if we want to have a 4th order, 2-pole bandpass filter with a Q of 5, the value of Q for both 2nd order, 1-pole sections must be 0.644 times the Q of the required 2-pole filter, or

$$Q_{1\text{-pole}} = 0.644\, Q_{2\text{-pole}} \qquad \text{(Eq. 8-22)}$$
$$= (0.644)(5)$$
$$= 3.22$$

By cascading three identical 2nd order, 1-pole sections, we then have a *6th order, 3-pole filter,* which has an ultimate rolloff of ±18 dB/octave, or ±60 dB/decade. The amplitude response is then expressed as

$$dB = 60 \log_{10}G - 30 \log_{10}\left[1 + Q^2\left(\frac{\omega^2 - 1}{\omega}\right)^2\right] \qquad \text{(Eq. 8-23)}$$

In a similar manner, cascading four identical 2nd order, 1-pole sections yields an *8th order, 4-pole filter,* with an ultimate rolloff of ±24 dB/octave, or ±80 dB/decade. For this filter, the amplitude response is

$$\text{dB} = 80 \log_{10}G - 40 \log_{10}\left[1 + Q^2\left(\frac{\omega^2 - 1}{\omega}\right)^2\right] \quad \text{(Eq. 8-24)}$$

Table 8-1. Multipliers To Be Used

Filter Type	Rolloff	$Q_{1\text{-pole}}$
4th order, 2-pole	±12 dB/octave	0.644 $Q_{2\text{-pole}}$
6th order, 3-pole	±18 dB/octave	0.510 $Q_{3\text{-pole}}$
8th order, 4-pole	±24 dB/octave	0.435 $Q_{4\text{-pole}}$

As an aid in determining the proper value of Q for the identical 1-pole sections to give the desired Q of the higher order/pole filter, the appropriate multipliers are presented in Table 8-1.

Example:

Design a 6th order, 3-pole multiple-feedback bandpass filter with a center frequency of 750 Hz, a passband gain of 6 (+15.6 dB), and a Q of 8.53.

Since we want to design a 6th order, 3-pole filter, *we must cascade three identical 2nd order, 1-pole filter sections.* From Table 8-1, the Q of each identical section must be

$$\begin{aligned} Q_{1\text{-pole}} &= 0.510\, Q_{3\text{-pole}} \\ &= (0.510)(8.53) \\ &= 4.35 \end{aligned}$$

In addition, since all three sections are to be identical, the voltage gain of all three sections must also be the same. For an overall voltage gain of 6, each section must have a gain of

$$\begin{aligned} G &= (6)^{1/3} \\ &= 1.82 \end{aligned}$$

All that is left is to design the required 1-pole section with $G = 1.82$ and $Q = 4.35$. Using the basic circuit of Fig. 8-1, we first choose a suitable value for C_2 and C_4, for example .033 μF. Then the three resistors are found, so that

$$\begin{aligned} R_1 &= 15.4\ \text{k}\Omega \quad \text{(Eq. 8-6)} \\ R_3 &= 777\Omega \quad \text{(Eq. 8-7)} \\ R_5 &= 55.9\ \text{k}\Omega \quad \text{(Eq. 8-9)} \end{aligned}$$

Using 5% resistors, the completed design is shown in Fig. 8-8.

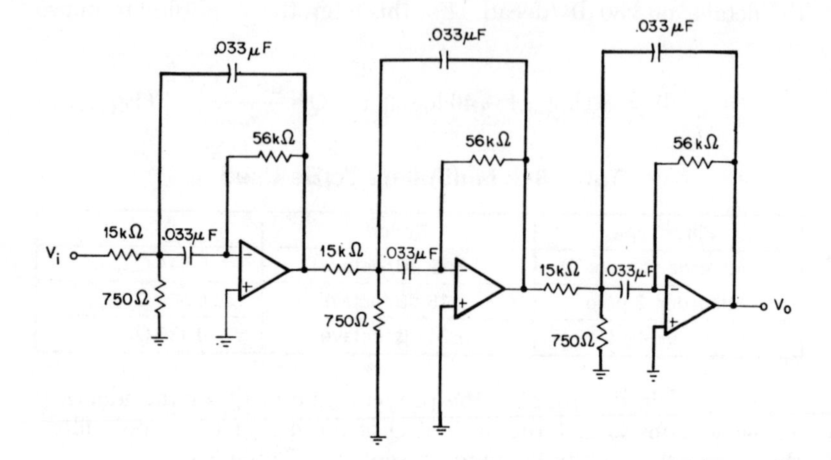

Fig. 8-8. A 6th order, 3-pole multiple-feedback bandpass filter.

The multiple-feedback bandpass filter, using a single op amp, is only useful for Qs less than approximately 15-20. However, with additional op amps, we can obtain Qs up to 50, but we run into serious problems that are beyond the scope of this book and will not be discussed here.* Nevertheless, a versatile high-Q bandpass filter, called a *state-variable filter,* is discussed in the next chapter.

Before concluding this section on bandpass filters, one general comment is in order. The multiple-feedback filter is not the only bandpass filter possible using a single op amp. Filter synthesis is an interesting game, and there are often many solutions for the same problem. Ask five electronic engineers to design a circuit to perform a given function and you will probably get five different solutions, which is particularly true about bandpass filter design. The multiple-feedback filter is only one of a number of possible bandpass filter networks using one operational amplifier, but is generally considered to be the best overall choice.

WIDEBAND FILTERS

Very often, particularly in audio applications, it is desired to pass a *wide band of frequencies* with relatively constant gain, as illustrated in Fig. 8-9. Such a bandpass response is said to be characteristic of a *wideband filter*. On the other hand the amplitude response

---

* See either: (1) Tobey, G. E., *et al, Operational Amplifiers—Design and Applications,* McGraw-Hill, 1971, pp. 299-303; or (2) Lancaster, D., *Active Filter Cookbook,* Howard W. Sams & Co., 1975, pp. 154-155.

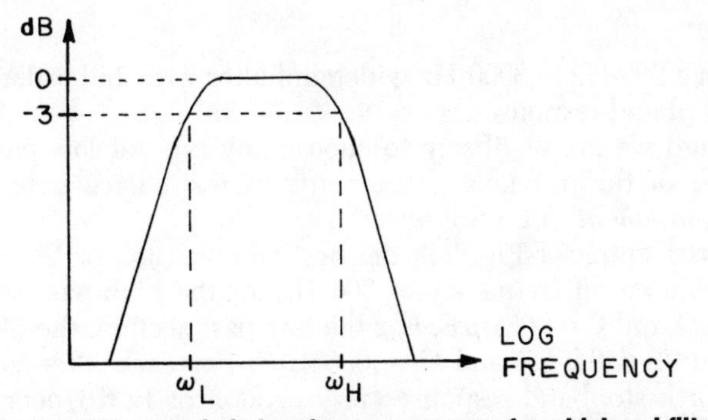

Fig. 8-9. Characteristic bandpass response of a wideband filter.

shown in Fig. 8-7 can be thought of as a *narrowband filter.* When is a bandpass filter a wideband filter? To my knowledge there is no set definition, but the following is probably as good as any:

If the 3-dB bandwidth of a bandpass filter is more than twice the center frequency, the filter is said to be wideband.

For most voice communications systems, it is desired to reject signals below 300 Hz and above 3000 Hz. The bandwidth is obviously 2700 Hz, and from Equation 3-4, the "center frequency," using this term only in the mathematical sense, is

$$f_o = [(300)(3{,}000)]^{1/2}$$
$$= 949 \text{ Hz}$$

and the Q is

$$Q = \frac{949 \text{ Hz}}{2700 \text{ Hz}}$$
$$= 0.35$$

For such a low value of Q it will be impractical, if not difficult, to design a multiple-feedback bandpass filter with the response similar to that shown in Fig. 8-9. To practically obtain such a wideband response, the best approach is to *cascade a low-pass filter, having a cutoff frequency of f_H, with a high-pass filter, having a cutoff frequency of f_L,* as shown in Fig. 8-10. Consequently, the low-pass and high-pass filters discussed in Chapters 5, 6, and 7 are ideal for this purpose.

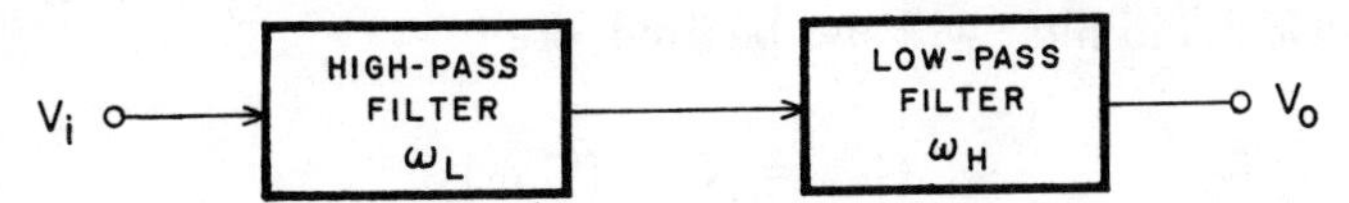

Fig. 8-10. Cascading a low-pass and high-pass filter to obtain wideband response.

Example:

Design a 300-Hz to 3000-Hz wideband filter with 2nd order Butterworth stopband responses.

Although we are at liberty to choose any type of low-pass/high-pass filter of the previous chapter, *the easiest choice is to use the "equal-component" VCVS filters.*

From the graph of Fig. 5-9, the best combination of 5% resistors that gives a cutoff frequency of 300 Hz for the high-pass section is R = 8.2 kΩ and C = .068 μF. For the low-pass section, the best combination is R = 8.2 kΩ and C = .0068 μF. For each filter to have a Butterworth stopband response (i.e., a rolloff of 12 dB/octave), the required input and feedback resistors are given in Table 7-2 (for 5% resistors), or Table 7-3 (for 1% resistors), resulting in the completed circuit of Fig. 8-11.

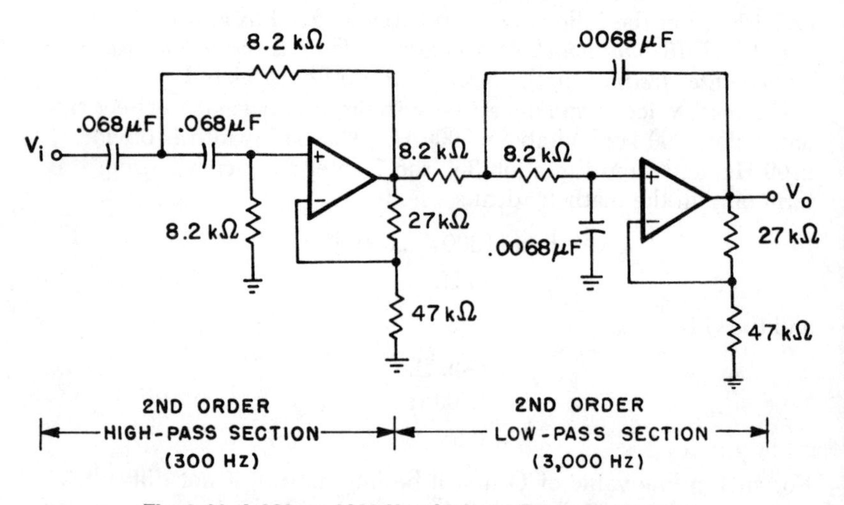

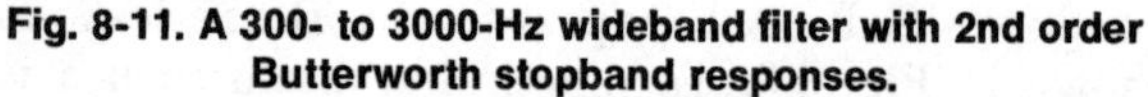

Fig. 8-11. A 300- to 3000-Hz wideband filter with 2nd order Butterworth stopband responses.

For this example, one concluding comment is in order. Since for this design, we used the "equal-component" VCVS filters, it should be kept in mind that the passband voltage gain of both sections is fixed at 1.586 (+4 dB). Consequently the overall passband gain for the wideband filter will also be fixed, or

$$
\begin{aligned}
G_{WB} &= (G_{LP})(G_{HP}) \\
&= (1.586)(1.586) \\
&= 3.17\ (+8\ \text{dB})
\end{aligned}
$$

If, for example, 5th order "equal-component" VCVS Butterworth filters were required, the passband gain of the wideband filter would be fixed at 6.60 (+16.4 dB) by using the value given in Table 7-3.

NOTCH FILTERS

The notch, or band reject filter, is often designed into audio and instrumentation systems for the rejection of a single frequency, such as 60-Hz power line frequency *hum.* Perhaps the best-known notch filter, although a *passive* filter, is the *twin-T* (sometimes spelled "twin-tee") filter, shown in Fig. 8-12.

For the twin-T circuit, the null frequency* is given by

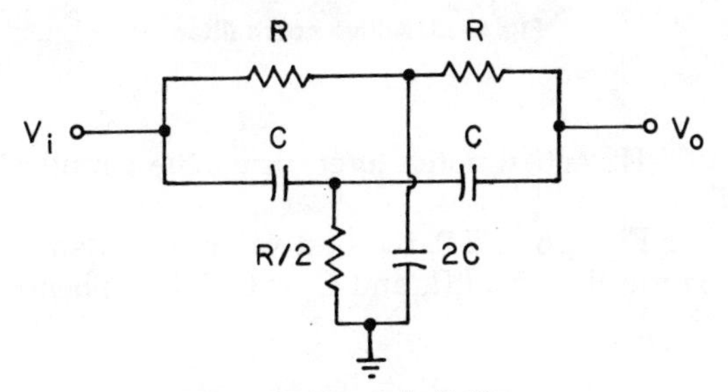

Fig. 8-12. The twin-T filter.

$$\omega_n = 1/RC \quad \text{(Eq. 8-25)}$$

or

$$f_n = 1/2\pi RC \quad \text{(Eq. 8-26)}$$

With almost perfect matching of the six components, the twin-T filter is theoretically capable of almost infinite rejection at the null frequency. However, using experimenter-grade components, you should expect the rejection, or *null depth,* to be only 30 to 40 dB.

Since this book is about *active filters,* we can place the twin-T network in an op-amp circuit, as shown in Fig. 8-13, to form an active notch filter.† The Q is found from

$$Q = R_2/2R_1 = C_1/C_2 \quad \text{(Eq. 8-27)}$$

Since the op-amp is basically connected as a voltage follower, the passband voltage gain is unity.

---

* When describing the notch filter, the terms *center frequency* and *null frequency* are often used interchangeably.

† Russel, H. T., "Design Active Filters With Less Effort," *Electronic Design,* January 7, 1971, pp. 82-85.

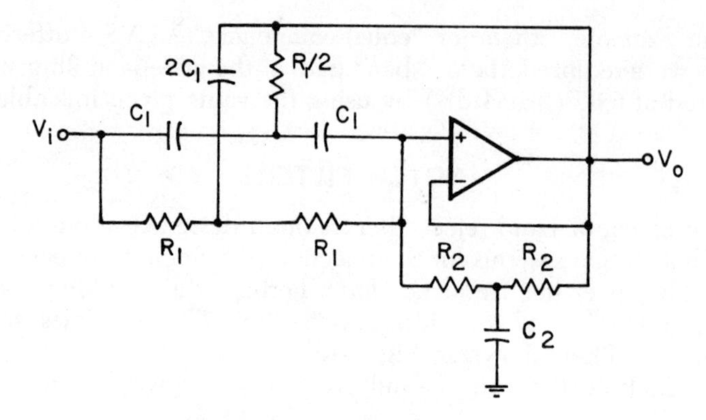

Fig. 8-13. Active notch filter.

Example:

Design a 60-Hz active notch filter, using the circuit of Fig. 8-13, with $Q = 5$.

From either Fig. 4-8 or 5-9, the best values of R_1 and C_1 to give a null at 60 Hz are $R_1 = 5.6\ k\Omega$, and $C_1 = 0.47\ \mu F$. Then by Equation 8-27,

$$R_2 = 56\ k\Omega$$

and

$$C_2 = 0.1\ \mu F$$

For the $R_1/2$ resistor, we simply use two 5.6 kΩ resistors connected *in parallel;* for the $2C_1$ capacitor, we use two .47-μF capacitors also connected *in parallel.* The completed design, using 5% resistors is shown below in Fig. 8-14.

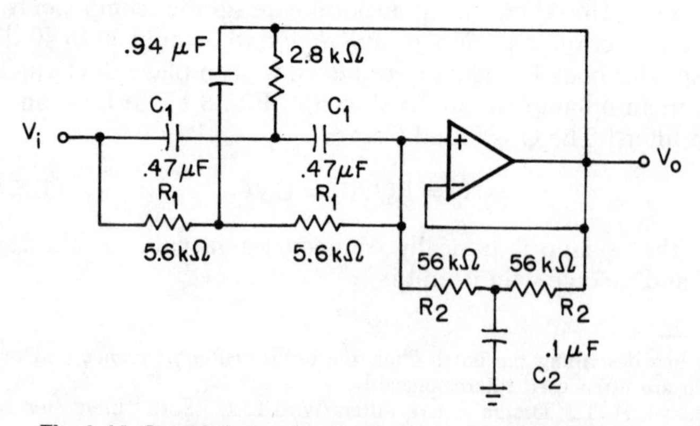

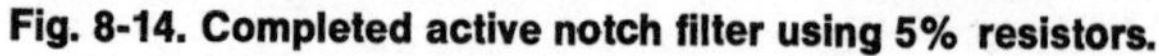
Fig. 8-14. Completed active notch filter using 5% resistors.

The placing of resistors in series and capacitors in parallel in the previous example was done to minimize the effect of the components' tolerance. However, close-tolerance components are not necessary in a notch filter if the null frequency can be adjusted. Using the network of Fig. 8-15,* the null frequency can be adjusted over a wide range by a single potentiometer. This type of network, referred to as a *bridged differentiator,* has a null frequency given by

$$f_n = \frac{1}{2\pi C[3R_1R_2]^{1/2}} \qquad \text{(Eq. 8-28)}$$

Naturally, the greatest rejection at the null frequency will result when the component values approach the exact theoretical design values. Consequently the three capacitors must be exactly equal, and the resistor that is connected from the input to output should be exactly six times the resistance in the variable resistance branch of the network. However this bridged-differentiator network will be considerably easier to adjust than the twin-T network, requiring only two potentiometers, as shown in Fig. 8-16.

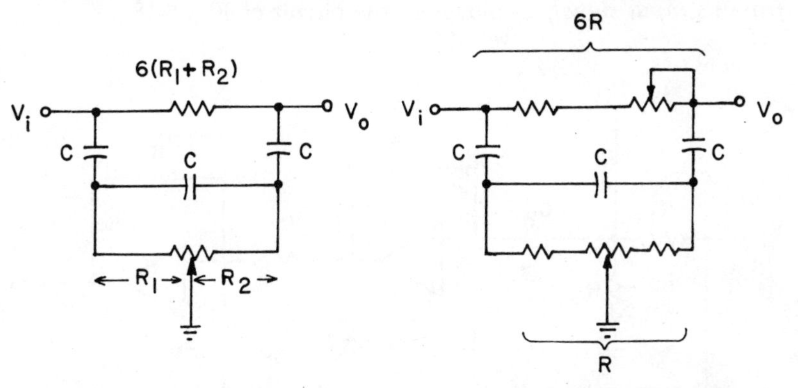

Fig. 8-15. A bridged differentiator.

Fig. 8-16. Bridged differentiator requires only two potentiometers to adjust.

An inexpensive active notch filter using the basic passive bridged-differentiator network, which can be tuned from 50 to 60 Hz with a minimum rejection of 30 dB, is shown in Fig. 8-17.† With 10% capacitors and 5% resistors, the bandwidth is approximately 14 Hz for a 50-Hz null frequency, and 18 Hz for 60 Hz.

---

* Hall, C., "Tunable RC Notch Filter," *Ham Radio,* September, 1975, pp. 16-20.

† Lefferson, P., "Tunable Notch Filter Suppresses Hum," *Electronics,* September 2, 1976, p. 100.

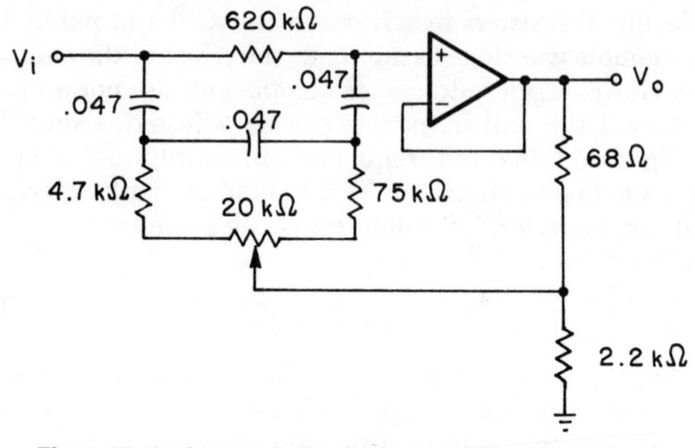

Fig. 8-17. Active notch filter using basic passive bridged-differentiator network.

Other active notch filters are formed so that one signal is subtracted from another. From classical filter theory, *a notch filter can be formed by subtracting the output signal from a bandpass filter from its input signal,* as shown by the circuit of Fig. 8-18.

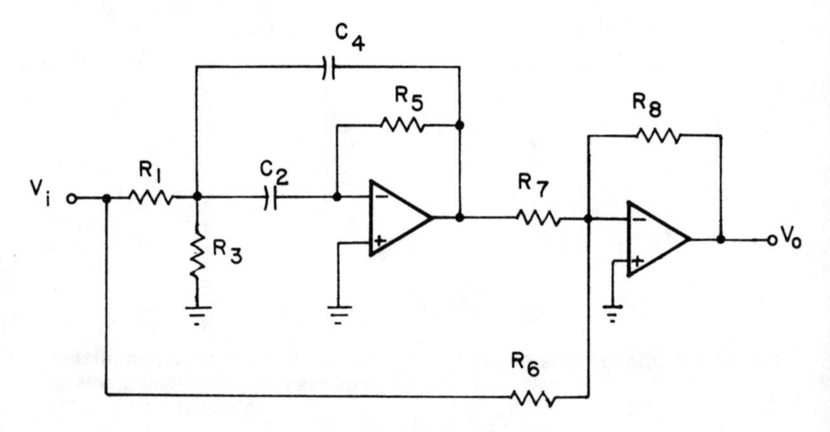

Fig. 8-18. Notch filter.

The first op-amp circuit is simply the basic multiple-feedback bandpass filter of Fig. 8-1, while the second op-amp acts as a summing amplifier. At the node of R_6 and R_7, the input signal is subtracted from the output signal of the bandpass filter section. To produce a deep null with this circuit, with C_2 and C_4 equal,

$$\frac{R_6}{R_7} = \frac{R_5}{2R_1} \quad \text{(Eq. 8-29)}$$

However, the right-hand term of the previous equation, $R_5/2R_1$, from Equation 8-9, is simply *the passband gain of the bandpass filter section.*

Example:

Design a 60-Hz notch filter using the circuit of Fig. 8-18. Assume that the passband gain is 3 and Q is 6 for the bandpass section.

Using Equations 8-6, 8-7, and 8-8, the design of the bandpass section is straightforward. Picking a value for C_2 and C_4, for example .22 μF, then

$$R_1 = \frac{(6)}{(3)(.22\ \mu F)(2\pi)(60\ Hz)} = 24.1\ k\Omega$$

$$R_3 = \frac{(6)}{[(2)(6)^2 - 3](.22\ \mu F)(2\pi)(60\ Hz)} = 1.05\ k\Omega$$

and

$$R_5 = \frac{(2)(6)}{(.22\ \mu F)(2\pi)(60\ Hz)} = 145\ k\Omega$$

Then from Equation, 8-29, R_6 and R_7 are found in terms of each other, or $R_6 = 3R_7$. If R_7 is 10 kΩ, then R_6 is 30 kΩ. The second op-amp section will have unity gain when $R_8 = R_7$, or 10 kΩ. Using 5% resistors, the completed circuit is shown in Fig. 8-19.

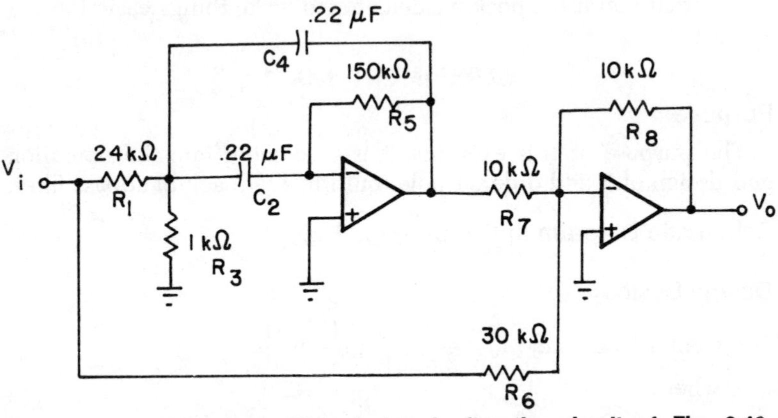

Fig. 8-19. A 60-Hz notch filter designed after the circuit of Fig. 8-18.

AN INTRODUCTION TO THE EXPERIMENTS

The following experiments are designed to demonstrate the design and operation of several types of bandpass and notch filters. With these experiments, we will have to be a little more careful in our measurements in order to accurately determine various parameters of interest.

The experiments that you will perform can be summarized as:

Experiment No.	*Purpose*
1	Demonstrates the operation and design of a multiple-feedback active filter.
2	Demonstrates the effect of cascading identical 2nd order, 1-pole bandpass sections.
3	Demonstrates the operation and design of a wideband filter, formed by cascading individual low-pass and high-pass filter sections.
4	Demonstrates the operation and design of a twin-T notch filter, using one op amp.
5	Demonstrates the operation and design of a notch filter, formed by a bandpass section in cascade with a summing amplifier.

NOTE: Because there will be a significant amount of numerical calculations in these experiments, it is strongly recommended that you use a pocket calculator to make things easier!

EXPERIMENT NO. 1

Purpose

The purpose of this experiment is to demonstrate the operation and design of a 2nd order, 1-pole multiple-feedback bandpass filter.

Schematic Diagram of Circuit (Fig. 8-20)

Design Basics

- Center frequency: $f_o = \frac{1}{2\pi C}\left[\frac{1}{R_5}\left(\frac{1}{R_1} + \frac{1}{R_3}\right)\right]^{1/2}$

  where

$$R_1 = \frac{Q}{2\pi f_o G C}$$

$$R_3 = \frac{Q}{2\pi f_o C(2Q^2 - G)}$$

$$R_5 = \frac{2Q}{2\pi f_o C}$$

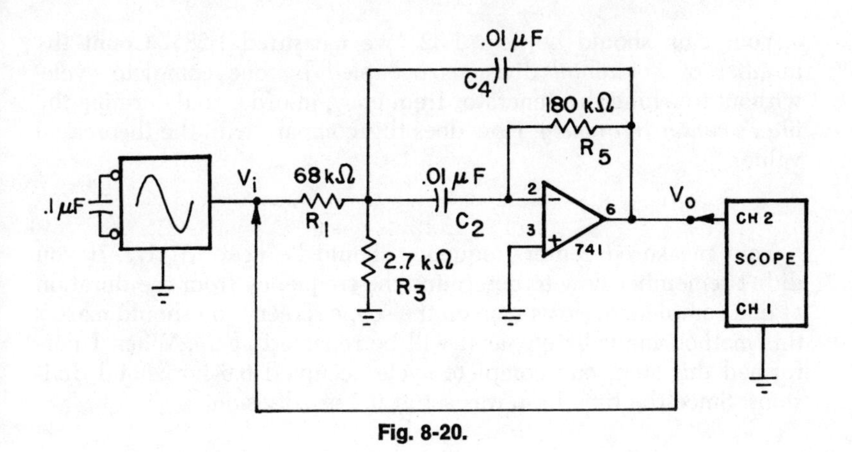

Fig. 8-20.

- Center frequency gain: $G = \dfrac{R_5}{2R_1}$
- Shifting of center frequency, keeping the passband gain and bandwidth constant:

$$R_3' = R_3\left[\frac{f_o}{f_o'}\right]^2$$

- Amplitude response:

$$20 \log_{10}G - 10 \log_{10}\left[1 + Q^2\left(\frac{f^2 - 1}{f}\right)^2\right]$$

Step 1

Set your oscilloscope for the following settings:

- Channels 1 & 2: 0.2 V/division
- Time base: 0.2 ms/division
- AC coupling

Step 2

Wire the circuit shown in the schematic diagram. Apply power to the breadboard and adjust the input voltage at 1.4 volts (i.e., 7 vertical divisions). Make this setting as accurate as you can!

Step 3

Now vary the generator frequency so that the output voltage, as displayed on Channel 2 of your oscilloscope, reaches its *maximum amplitude*. Measure this output voltage and determine the voltage gain at this point. It may be necessary to reduce the sensitivity of Channel 2 to 0.5 V/ division. How does your measured gain compare with the expected value?

Your gain should be near 1.32 (we measured 1.28). Count the number of *horizontal* divisions occupied by one complete cycle, without touching the generator frequency, in order to determine the filter's *center frequency*. How does this compare with the theoretical value?

Your measured center frequency should be near 737 Hz. If you didn't remember how to determine the frequency from the duration of one complete cycle as seen on the scope screen, you should master this method immediately, as it will be repeated again. When I performed this step, one complete cycle occupied 6.6 horizontal divisions. Since the time base was set at 0.2 ms/division,

$$f = \frac{1}{(6.6 \text{ divisions}) \times (0.2 \text{ ms/division})}$$

$$= \frac{1}{1.32 \text{ ms}}$$

$$= 758 \text{ Hz}$$

Step 4

Now determine the upper and lower 3-dB frequencies by measuring the frequencies at which the amplitude response drops by 0.707 times the maximum voltage gain, which is the same as a 3-dB decrease. To do this, multiply your measured center frequency gain by 0.707 (in our case, this value is 0.90). Then multiply this intermediate value by 1.4 (i.e., the filter's input voltage) to obtain the output voltage at which the response is 3 dB less than the center frequency gain (e.g., in our case, 1.27 volts). After you have determined this value, what are the upper and lower 3-dB frequencies (to be found by varying the generator frequency above and below the center frequency until the peak-to-peak voltage reaches the particular value you have just calculated (e.g., 1.27 volts).

f_L = ______ Hz

f_H = ______ Hz

Step 5

Subtract the lower value from the higher value (the *3-dB bandwidth*) and divide this value into the value you determined as the center frequency (Step 3), which then is the Q, or *quality factor*,

$$Q = \frac{f_o}{f_H - f_L}$$

$Q =$ ______

You should have determined an approximate Q of 4.17. If not, repeat Steps 3 and 4, carefully measuring the voltages and frequencies! In our case, the measured 3-dB frequencies were 672 and 860 Hz, so that the bandwidth was 188 Hz, and the filter's Q was 758/188 or 4.03.

Step 6

From our measured 3-dB frequencies, we can check whether Equation 3-4 holds, since the center frequency, by definition, *is the geometric average of the two 3-dB frequencies, or*

$$f_o = (f_H f_L)^{1/2}$$

Using the above equation, does your result reasonably compare with the value you determined in Step 4?

Step 7

Disconnect the power to the breadboard and replace the 2.7-kΩ resistor (R_3) with a 1.5-kΩ resistor. Reconnect the power to the breadboard, and adjust the input voltage now at 1.00 volt peak-to-peak. Now repeat Steps 3, 4, 5, and 6 to determine the filter's voltage gain, center frequency, bandwidth, and Q, completing the following table:

Center frequency gain = ______

Upper 3 dB frequency = ______ Hz

Lower 3 dB frequency = ______ Hz

Bandwidth = ______ Hz

Center frequency = ______ Hz

$Q =$ ______

Step 8

By changing the value of resistor R_3, does the new center frequency that you determined in the previous step compare favorably with the equation given in the "design basics" section?

From the equation given in the "design basics" section, the new center frequency should be approximately

$$f_o' = (758\ \text{Hz})\left[\frac{2.7\ \text{k}\Omega}{1.5\ \text{k}\Omega}\right]^{1/2}$$

$$= 1017\ \text{Hz}$$

which is based on the center frequency that you determined in Step 3.

Step 9

If you have done everything carefully, you should find that the bandwidth, as found in Step 7, should be approximately the same as the bandwidth of the original circuit, even though the center frequency has been shifted. Consequently, the changing of a single resistor changes only the center frequency. The filter's bandwidth and center frequency gain remain the same. If you are still not convinced, try another resistance for R_3 and repeat Steps 3 through 7.

Step 10

Disconnect the power to the breadboard and replace the 2.7-kΩ resistor for R_3, as in the original schematic diagram. Then, after reconnecting power, make sure that the input voltage is 1.4 volts peak-to-peak, and adjust the generator frequency at 100 Hz (the scope's time base should be set to 1 ms/division so that the signal's period will occupy exactly 10 horizontal divisions) to complete the following short table.

Frequency	V_o	V_o/V_i	Experimental dB Gain
100 Hz			
200 Hz			

The difference in the above two frequencies is 1 octave, which is far enough away from the filter's center frequency to determine the ultimate rolloff of a 2nd order, 1-pole bandpass filter. Subtract the dB gain measured at 200 Hz from the dB gain at 100 Hz. What do you get?

You should find that the rolloff of a 2nd order, 1-pole bandpass filter is nearly 6 dB/octave. Of course, since the bandpass filter's amplitude response is *symmetrical about the center frequency,* we should find the same result for a 1 octave frequency difference sufficiently above the filter's center frequency, as will be shown in the next step.

Step 11

Set the oscilloscope time base to .1 ms/division and then adjust the generator frequency to 5000 Hz (5 complete cycles for the 10 horizontal divisions), and determine the dB gain. Then adjust the frequency at 10,000 Hz and also determine the dB gain, completing the following table.

Frequency	V_o	V_o/V_i	Experimental dB Gain
5000 Hz			
10,000 Hz			

For this case, what is the rolloff?

The rolloff should also be approximately 6 dB/octave. Since the filter's response is symmetrical about the center frequency, the rolloff above and below the center frequency for a 2nd order, 1-pole bandpass filter will be 6 dB/octave or 20 dB/decade.

Step 12

As an optional exercise, remove the 2.7-kΩ resistor from the circuit, and then determine the filter's center frequency, gain, bandwidth, and Q. Compare your results with Equations 8-12, 8-13, and 8-15.

EXPERIMENT NO. 2

Purpose

The purpose of this experiment is to demonstrate the effect of cascading two identical 2nd order, 1-pole bandpass filter sections.

Schematic Diagram of Circuit (Fig. 8-21)

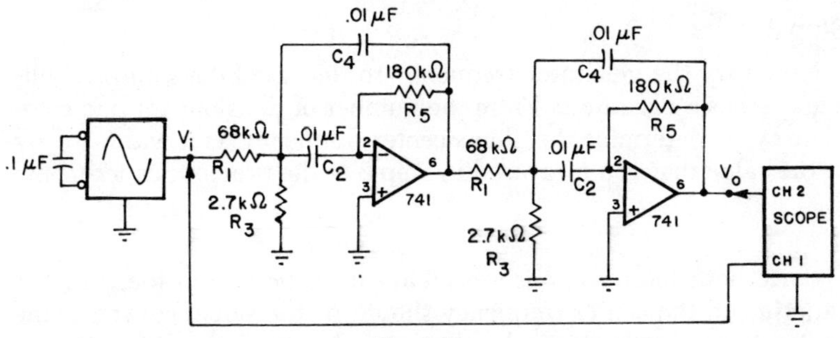

Fig. 8-21.

Design Basics

- Center frequency: $f_o = \frac{1}{2\pi C}\left[\frac{1}{R_5}\left(\frac{1}{R_1} + \frac{1}{R_3}\right)\right]^{1/2}$

where
$$R_1 = \frac{Q}{2\pi f_o GC}$$
$$R_3 = \frac{Q}{2\pi f_o C(2Q^2 - G)}$$
$$R_5 = \frac{2Q}{2\pi f_o C}$$

- Center frequency gain (each section):

$$G = \frac{R_5}{2R_1}$$

- Overall center frequency gain: G^2
- Overall Q of the resultant 4th order, 2-pole filter:

$$Q_{2\text{-pole}} = 1.553\ Q_{1\text{-pole}}$$

- Amplitude response:

$$40 \log_{10}G - 20 \log_{10}\left[1 + Q_{2\text{-pole}}{}^2\left(\frac{f^2 - 1}{f}\right)^2\right]$$

Step 1

Set your oscilloscope for the following settings:

- Channels 1 & 2: 0.2 V/division
- Time base: 0.2 ms/division
- AC coupling

Step 2

Wire the circuit shown in the schematic diagram. Apply power to the breadboard and adjust the input voltage at 0.8 volt peak-to-peak.

Step 3

Now vary the generator frequency so that the filter's output voltage reaches a maximum. From the number of divisions for one complete cycle, determine the filter's center frequency. Is it within 10 Hz of the value that you measured in Step 3 of the previous experiment?

Since both filter sections are identical to the one in the previous experiment, the center frequency should be the same. However, the center frequency may not be the same due to slight differences in the component values of the two sections, especially the .01-μF capacitors. By looking at the schematic diagram, you should be able to see the importance of having the capacitors equal, since there are 4 of them, all .01 μF! If the center frequency is more than 10 Hz difference, try alternate .01-μF capacitors.

Step 4

At this center frequency, determine the gain. Is it what you would expect?

Assuming that both filter sections are nearly identical to the one of the previous experiment, the center frequency gain should equal the square of the gain of a single section. When I performed this experiment, the overall gain measured was 1.62. Since the gain of the first section (determined in Experiment No. 1) was 1.28, then the gain of the second section must be 1.27. This slight difference is not much to be concerned about.

On the other hand, if we went to the trouble of "matching" the resistors with a digital ohmmeter, and the capacitors with a capacitance bridge, then we should be concerned if there are significant differences between the two sections.

Step 5

Now carefully determine the upper and lower 3 dB frequencies by measuring the frequencies at which the output voltage (and also the voltage gain) decreases by a factor of 0.707. In our experiment, the output voltage at the 3-dB frequencies were determined to be, for example:

$$\begin{aligned} V_o(3\text{ dB}) &= V_o(\text{center frequency}) \times 0.707 \\ &= 1.30 \times 0.707 \\ &= 0.92 \text{ volts} \end{aligned}$$

Writing your results below:

$f_L =$ _______ Hz

$f_H =$ _______ Hz

determine the bandwidth, and Q. How do your results compare with the theoretical values?

Since the theoretical Q for each section is 4.17, the overall Q is 1.553 times this value, or 6.48.

Step 6

Now adjust the oscilloscope time base at 1 ms/division. Although the sensitivity of Channel 1 will be kept at 0.2 V/division, now change the sensitivity of Channel 2 to 10 mV/division. Adjust the input voltage at 1.0 volt peak-to-peak, and the input frequency at 200 Hz, in order to complete the following table.

Frequency	V_o	V_o/V_i	Experimental dB Gain
200 Hz			
400 Hz			

Subtract the dB gain measured at 200 Hz from the dB gain measured at 400 Hz. What do you get?

You should have found that this difference is approximately +12 dB/octave, which is the rolloff for a 4th order, 2-pole bandpass filter below the center frequency.

Step 7

Since the response of a bandpass filter is symmetrical about its center frequency, we should find the rolloff for a 1 octave difference above the center frequency. Now set the oscilloscope time base at .1 ms/division, and the generator frequency at 2000 Hz in order to complete the following table. Remember, the input voltage is still 1.0 volt peak-to-peak.

Frequency	V_o	V_o/V_i	Experimental dB Gain
2000 Hz			
4000 Hz			

Subtract the dB gain measured at 2000 Hz from the dB gain measured at 4000 Hz. How does your answer compare with the one found in Step 6?

This difference should be approximately −12 dB/octave, and is the same as in Step 6, except that the rolloff is *negative,* since we have just looked at a 1 octave difference *above the center frequency.*

EXPERIMENT NO. 3

Purpose

The purpose of this experiment is to demonstrate the operation and design of a wideband passband active filter, formed by cascading individual 2nd order low-pass and high-pass Butterworth filter sections.

Schematic Diagram of Circuit (Fig. 8-22)

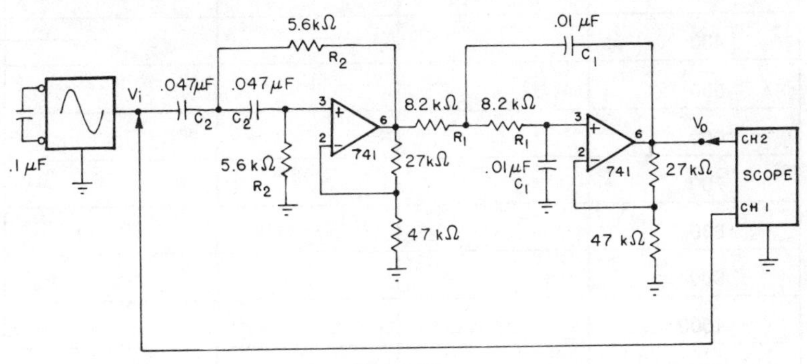

Fig. 8-22.

Design Basics

- Low-pass cutoff frequency: $f_1 = 1/2\pi R_1C_1$
- High-pass cutoff frequency: $f_2 = 1/2\pi R_2C_2$
  where f_1 is greater than f_2
- Passband gain (both sections): 1.586 (+4 dB) each
- Maximum filter gain: 2.52 (+8 dB)

Step 1

Set your oscilloscope for the following settings:

- Channel 1: 200 mV/division
- Channel 2: 50 mV/division
- Time base: 1 ms/division
- AC coupling

Step 2

Wire the circuit shown in the schematic diagram. Apply power to the breadboard and adjust the input voltage at 1.0 volt peak-to-peak and the generator frequency at 100 Hz.

Step 3

Starting at 100 Hz, vary the generator frequency and complete the following table, plotting your results on the blank graph provided for this purpose in Fig. 8-23. Remember, keep the input voltage level constant at 1.0 volt over the entire range!

Frequency	V_o	V_o/V_i	Experimental dB Gain
100 Hz			
200			
300			
400			
500			
600			
700			
800			
900			
1000			
2000			
3000			
4000			
5000			

Step 4

From your results, where are the low-pass and high-pass cutoff frequencies?

You should find that frequencies that are 3 dB less than the maximum response should be approximately 600 Hz (high-pass) and 2000 Hz (low-pass). For the component values used in this experiment, the actual frequencies are 605 Hz and 1942 Hz respectively.

Step 5

How does the maximum voltage gain that you determined in this experiment compare with the expected value?

Although the maximum voltage gain, at approximately 1100 Hz, is 2.52 (+8 dB), your measured value should be a little less than this

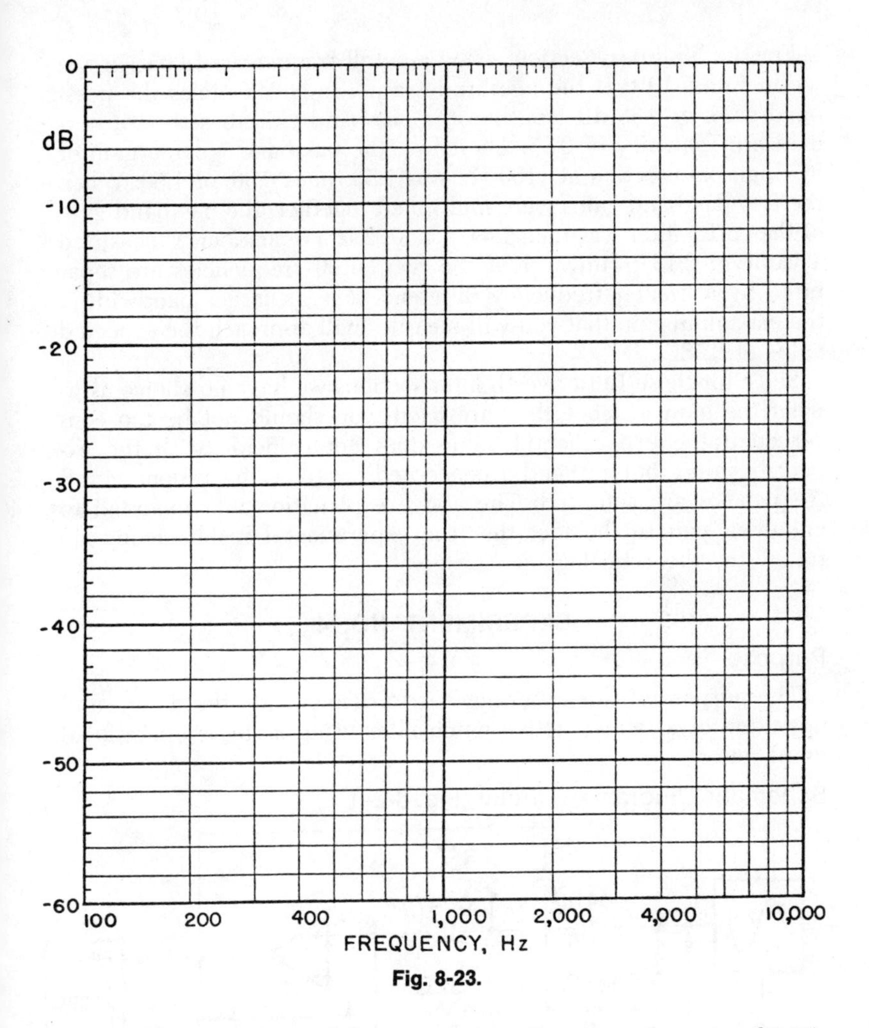

Fig. 8-23.

value. When I performed this experiment, I measured a gain of 2.15. There is a simple explanation why this is so.

First, I found that the passband gain for the *low-pass* section was 1.52, and the passband gain for the *high-pass* section was 1.56. Consequently, the maximum possible gain that could be expected is 2.37. However, if we refer to the graph of Fig. 5-2, which is the normalized amplitude response for a 2nd order low-pass Butterworth filter, the amplitude response at approximately 1100 Hz is 0.43 dB less than the passband gain (or 0.95 times the maximum value), based on a cutoff frequency of 1942 Hz (a normalized frequency of 0.57). Therefore, the passband voltage gain of my low-pass section at 1100 Hz was 1.52 times 0.95, or 1.44.

For the high-pass section, using a cutoff frequency of 605 Hz, we see in Fig. 5-10 that the amplitude response is 0.96 times the passband gain, or 0.38 dB less than the passband gain, based on a normalized frequency of 0.55. Therefore, the passband voltage gain of my high-pass section at 1100 Hz was 1.56 times 0.96, or 1.50. When the two passband gains were multiplied together, the passband gain of the entire filter was then 1.44 × 1.50, 2.16 (remember, I measured a value of 2.15). However, if the two cutoff frequencies are separated by a greater frequency difference (i.e., a larger bandwidth) the maximum gain that you will measure will approach the expected value of 2.52.

Since for these Butterworth filter sections we have no choice as to what the gain is, since they are fixed, you should not be too concerned if the experimental value does not coincide with the expected value, but rather be concerned whether the proper cutoff frequencies are achieved. The above explanation was included to enlighten you to the fact that the experimental results really do follow the theoretical.

EXPERIMENT NO. 4

Purpose

The purpose of this experiment is to demonstrate the design and operation of a twin-T active notch filter using a single operational amplifier.

Schematic Diagram of Circuit (Fig. 8-24)

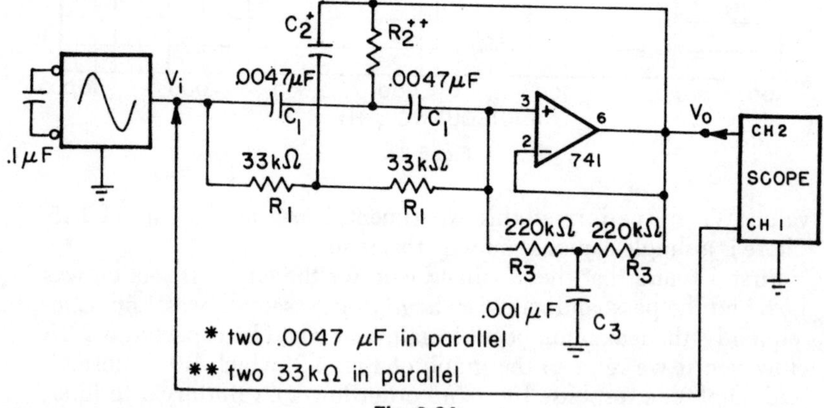

Fig. 8-24.

Design Basics

- Notch frequency: $f_n = 1/2\pi R_1 C_1$
- $R_1 = 2 R_2$

- $C_2 = 2\,C_1$
- $R_3 = 4\,QR_2$
- $C_3 = \dfrac{C_2}{2Q}$
- Passband gain: unity

Step 1

Set your oscilloscope for the following settings:

- Channels 1 & 2: 0.2 V/division
- Time base: 1 ms/division
- AC coupling

Step 2

Wire the circuit shown in the schematic diagram. Apply power to the breadboard and adjust the input voltage at 1.4 volts peak-to-peak (7 vertical divisions) and the frequency at 100 Hz. The output voltage should also be 1.4 volts peak-to-peak, since the passband gain is unity.

Step 3

Next, slowly increase the input frequency until the output voltage, as seen on Channel 2 of the oscilloscope reaches a minimum value, which is the filter's *null frequency*. Measure the output voltage and determine the voltage gain, or *depth of null*, at this null frequency and record your result below:

depth of null = ______ dB

For this circuit, the depth of the null will depend on how closely the relationships between R_1 and R_2, and C_1 and C_2 hold. As a comparison, when I performed this experiment, I obtained a null depth of 21.5 dB.

Step 4

Now measure the null frequency and record your result below:

f_n = ______ Hz

How does the value compare with the expected null frequency?

You should have determined a null frequency near the calculated frequency of 1027 Hz. For my circuit, the frequency was 1053 Hz.

Step 5

Vary the frequency above and below the null frequency, noting the two frequencies at which the output voltage equals 1.0 volt.

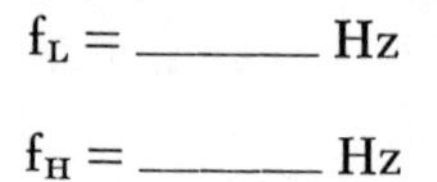

$$f_L = \_\_\_\_\_\_ \text{ Hz}$$

$$f_H = \_\_\_\_\_\_ \text{ Hz}$$

From the above two frequencies, and the null frequency found in Step 4, determine the filter's Q. How does it compare with the intended value of 3.33?

EXPERIMENT NO. 5

Purpose

The purpose of this experiment is to demonstrate the operation and design of an active notch filter made from cascading a multiple-feedback bandpass filter section with a summing amplifier.

Schematic Diagram of Circuit (Fig. 8-25)

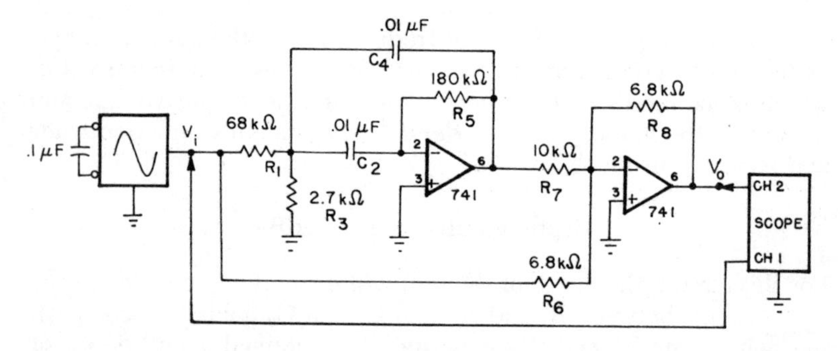

Fig. 8-25.

Design Basics

- Bandpass filter section
  - Center frequency: $f_n = \frac{1}{2\pi C}\left[\frac{1}{R_5}\left(\frac{1}{R_1} + \frac{1}{R_3}\right)\right]^{1/2}$
  - Center frequency gain: $G = \frac{R_5}{2R_1}$
- Summing amplifier section
  - For unity passband gain: $R_6 = R_8$ and $R_7 = GR_8$ where G is the passband gain of the bandpass filter section

Step 1

Set your oscilloscope for the following settings:

- Channels 1 & 2: 0.2 V/division
- Time base: 1 ms/division
- AC coupling

Step 2

Wire the circuit shown in the schematic diagram. Apply power to the breadboard and adjust the input voltage at 1.4 volts, and the frequency at 100 Hz. What is the voltage gain of this circuit at this frequency?

You should have measured an output voltage that is approximately equal to the input, so that the filter's gain (essentially the passband gain) is 1, or unity.

Step 3

Now set Channel 2 at 10 mV/division and the time base at 0.2 ms/division. Vary the generator frequency so that the output voltage, as displayed on Channel 2 of your oscilloscope reaches a *minimum amplitude*. Measure the output voltage and frequency to determine the *null depth in dB* at this null frequency. How does your value for the null frequency compare with the center frequency that you determined in Step 3 of Experiment No. 1?

Since the bandpass section for this experiment is exactly the same as the circuit for the first experiment of this chapter, the null frequency of this experiment should be nearly the same as the center frequency of Experiment No. 1. When I performed this experiment, my null frequency was 794 Hz, compared with a center frequency of 758 Hz for Experiment No. 1. However, I discovered that I didn't use the same .01-μF capacitors in both experiments! Assuming that you have used the same capacitors in both experiments, the two frequencies should be approximately the same, within 5 Hz. The null depth that I measured (the dB gain at the null frequency) was −25.9 dB, as a comparison. For this circuit, you should obtain a null depth of at least −25 dB without using "matched" components.

Step 4

Now vary the generator frequency above and below this null frequency to determine the filter's 3-dB frequencies. This is found by

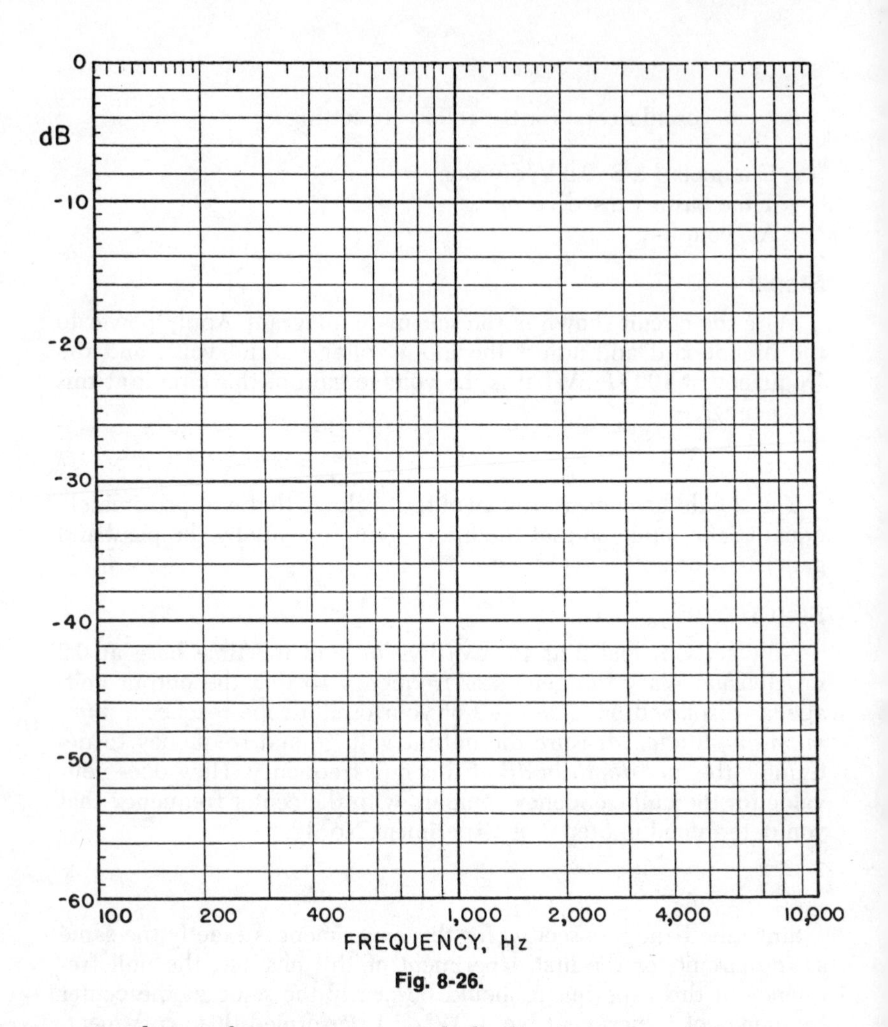

Fig. 8-26.

noting the two frequencies at which the filter's output voltage equals 1 volt peak-to-peak (i.e., $1.4 \times 0.707 = 1.0$). Write your results below, and subtract the two frequencies to determine the filter's 3-dB bandwidth.

$f_H =$ ______ Hz

$f_L =$ ______ Hz

3-dB bandwidth = ______ Hz

If nothing else, does the bandwidth, as determined above, nearly equal the bandwidth determined in Step 5 of Experiment No. 1?

The two bandwidths should be equal, to within 5 Hz. If, for example, we are able to measure the upper and lower 3-dB frequencies better than the filter's null frequency, calculate the null frequency from the measured 3-dB frequencies using the formula,

$$f_n = (f_L f_H)^{1/2}$$

Does the calculated result approximately equal the result you obtained in Step 3?

Step 5

To demonstrate the shape of the notch filter's amplitude response, vary the generator frequency from 100 Hz to approximately 5000 Hz, plotting your results on the blank graph provided for this purpose in Fig. 8-26.

CHAPTER 9

The State-Variable Filter

INTRODUCTION

In this chapter, a different type of multiple feedback filter, called the state-variable filter, will be presented. Although this type of filter uses at least three operational amplifiers, we nevertheless have the capability of having simultaneous low-pass, high-pass, and bandpass output responses. When used with an additional op amp, we can use the state-variable filter to form a notch filter.

OBJECTIVES

At the completion of this chapter, you will be able to do the following:

- Design state-variable filters with simultaneous low-pass, high-pass, and bandpass output responses
- Combine two outputs of a state-variable filter to form a notch filter

THE STATE-VARIABLE FILTER

As shown in Fig. 9-1, the state-variable filter, sometimes called a *universal filter,* is basically made up of several different functions:

- 1 summing block
- 2 identical integrators
- 1 damping network

Because of the manner that these functions are interconnected, we are able to *simultaneously* have the following filter responses:

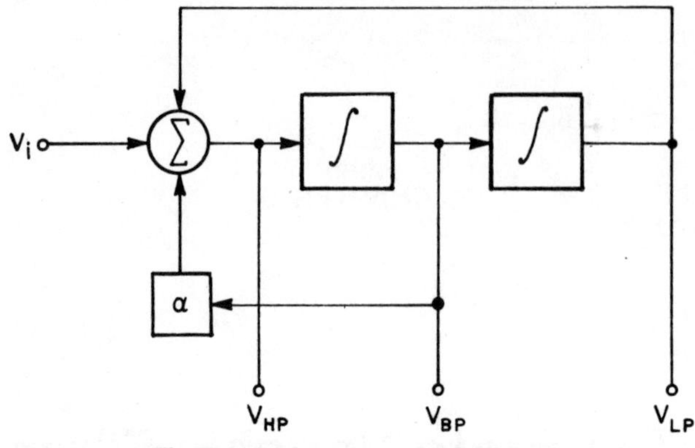

Fig. 9-1. Basic state-variable filter.

- A 2nd order low-pass filter
- A 2nd order high-pass filter
- A 1-pole bandpass filter

The cutoff frequency of the low-pass and high-pass responses *is identical* to the center frequency of the bandpass response. In addition, the damping factor α (equal to $1/Q$ for the bandpass filter) is the same *for all three responses.*

THE UNITY-GAIN STATE-VARIABLE FILTER

When the diagram of Fig. 9-1 is replaced by a circuit having three op amps, we have the *unity-gain state-variable filter** shown in Fig. 9-2, which has been normalized for a cutoff frequency of 1 radian/s and an impedance level of 1Ω. The first op amp A_1 is the *summing block* for the input, low-pass, and bandpass signals, which is the same circuit shown in Fig. 2-9. In series with this summing amplifier are two *identical op-amp integrators* (A_2 and A_3)† which determine the cutoff and center frequency by the formulas

$$\omega_c \text{ or } \omega_o = 1/RC \text{ radians/s} \qquad \text{(Eq. 9-1)}$$

or

$$f_c \text{ or } f_o = 1/2\pi RC \text{ Hz} \qquad \text{(Eq. 9-2)}$$

The damping network is composed of resistors R_A and R_B. For both the low-pass and high-pass responses to have a *2nd order Butterworth response,* the damping factor must equal 1.414, as pointed out in Chapters 5 and 6. Therefore, R_A is equal to 1.12Ω when R_B is 1Ω. For minimum offset (i.e., the output voltage when the input

---

* A complete analysis of the state-variable filter is given in the Appendix.

† Refer to Chapter 2.

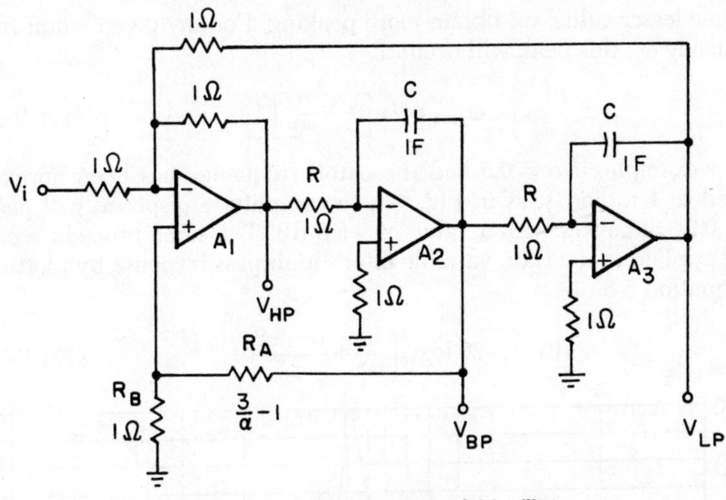

Fig. 9-2. A unity-gain state-variable filter.

voltage is zero), R_B should be 0.33Ω, which is the parallel combination of the two 1Ω input resistors connected to the *inverting input* of A_1, and the 1Ω feedback resistor. Then R_A should be 0.37Ω. However, if the offset voltage can be neglected, the following relationship must hold:

$$R_A = \left[\frac{3}{\alpha} - 1\right] R_B \qquad \text{(Eq. 9-3)}$$

or, since $Q = 1/\alpha$

$$R_A = (3Q - 1)R_B \qquad \text{(Eq. 9-4)}$$

In addition, if the offset of the integrators can be neglected, the resistors connected to the *noninverting inputs* of A_2 and A_3 may be replaced by short circuits to ground.

However, when we make the damping factor equal to 1.414 for a 2nd order Butterworth low-pass or high-pass response, the response of the bandpass filter suffers terribly, since the bandpass Q is now 0.707! If we let Q = 10, for example, then the damping factor for the low-pass and high-pass responses is 0.1, which is definitely not a Butterworth response.

By plotting Equation 5-3 for unity passband gain,

$$\text{dB} = -20 \log_{10}[\omega^4 + (\alpha^2 - 2)\omega^2 + 1]^{1/2} \qquad \text{(Eq. 9-5)}$$

and damping factors varying from 0.1 to 1.414 (i.e., Q varies from 10 to 0.707), we see in Fig. 9-3 that a *peaking* occurs near the cutoff frequency for the low-pass response of the state-variable filter. As we go from $\alpha = 1.414$ (a maximally flat, or Butterworth response) to some lesser value, we obtain more peaking. For any given cutoff frequency ω_c, this peak will occur at

$$\omega_{\text{peak}} = \omega_c \left[1 - \frac{\alpha^2}{2}\right]^{1/2} \qquad \text{(Eq. 9-6)}$$

For example, if $\alpha = 0.5$ and the cutoff frequency has been normalized to 1 radian/s, as in Fig. 9-3, the amplitude response will peak at 0.94 radians/s with a value of +6.3 dB. The same procedure can be applied to the state-variable filter's high-pass response by plotting Equation 5-8,

$$\text{dB} = -20 \log_{10}\left[\frac{1}{\omega^4} + \frac{\alpha^2 - 2}{\omega^2} + 1\right]^{1/2} \qquad \text{(Eq. 9-7)}$$

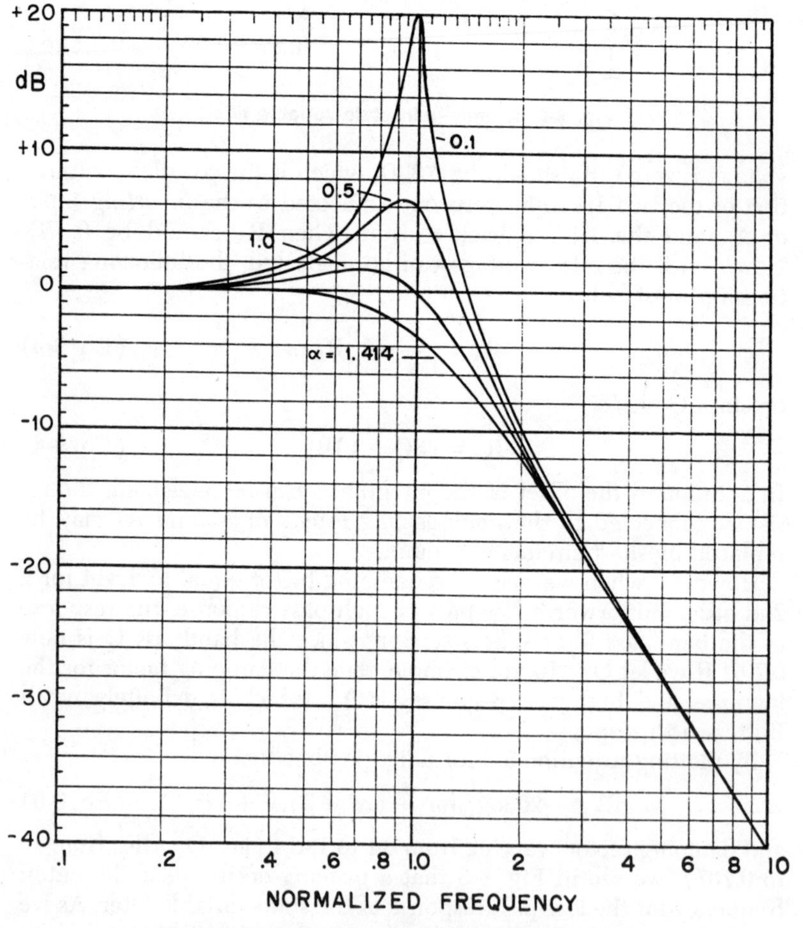

Fig. 9-3. Low-pass response of circuit in Fig. 9-2.

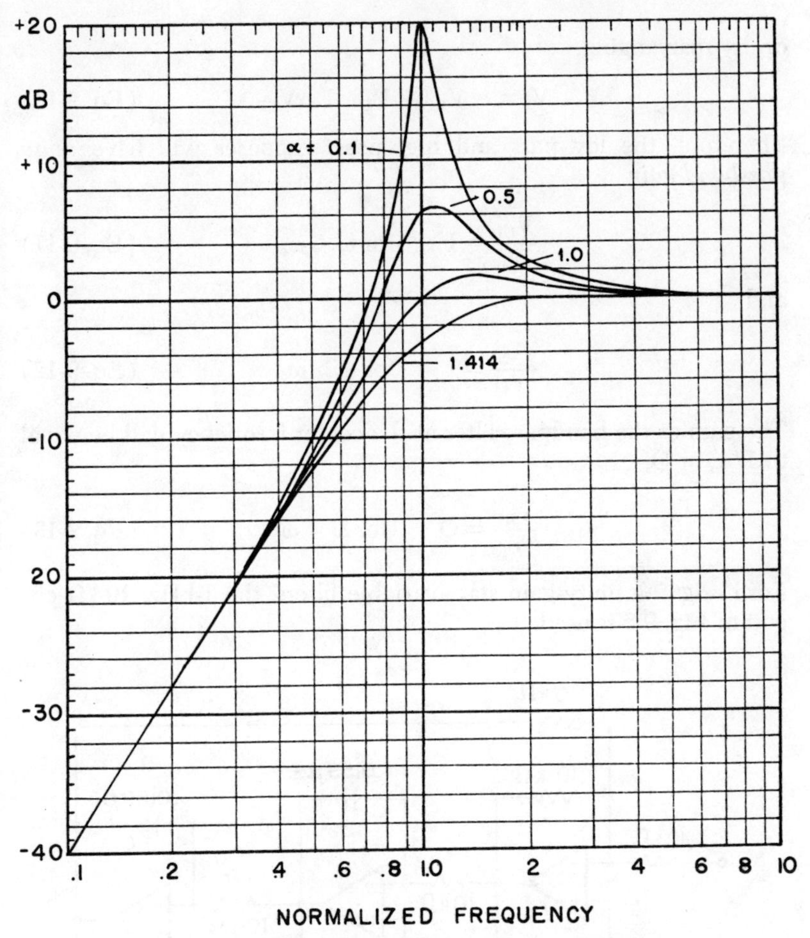

Fig. 9-4. High-pass response of circuit in Fig. 9-2.

as shown in Fig. 9-4. The peak response will occur at

$$\omega_{\text{peak}} = \omega_c / \left(1 - \frac{\alpha^2}{2}\right)^{1/2} \qquad \text{(Eq. 9-8)}$$

Consequently, *it is not possible to obtain the optimum performance with all three outputs, and we have to compromise.* We then should design either for a 2nd order Butterworth low-pass/high-pass response or for a high-Q bandpass response.

Comparing the summing amplifier of Fig. 9-2 with that of Fig. 2-9, we can obtain interesting results. Using Equation 2-12, the *high-pass output* V_{HP} can be written as

$$V_{HP} = -V_i - V_{LP} + \alpha V_{BP} \qquad \text{(Eq. 9-9)}$$

or, by rearranging,

$$V_i = -V_{LP} - V_{HP} + \alpha V_{BP} \qquad \text{(Eq. 9-10)}$$

Therefore, the low-pass and high-pass responses will have *unity passband gain*

$$\frac{V_{LP}}{V_i} = 1 \qquad \text{for } \omega < \omega_c \qquad \text{(Eq. 9-11)}$$

and

$$\frac{V_{HP}}{V_i} = 1 \qquad \text{for } \omega > \omega_c \qquad \text{(Eq. 9-12)}$$

The gain of the bandpass filter at the center frequency will be equal to $1/\alpha$, or Q,

$$\frac{V_{BP}}{V_i} = Q \qquad \text{for } \omega = \omega_o \qquad \text{(Eq. 9-13)}$$

For *designing* unity-gain state-variable filters, the 1-kHz, 10-kΩ circuit of Fig. 9-5 is used.

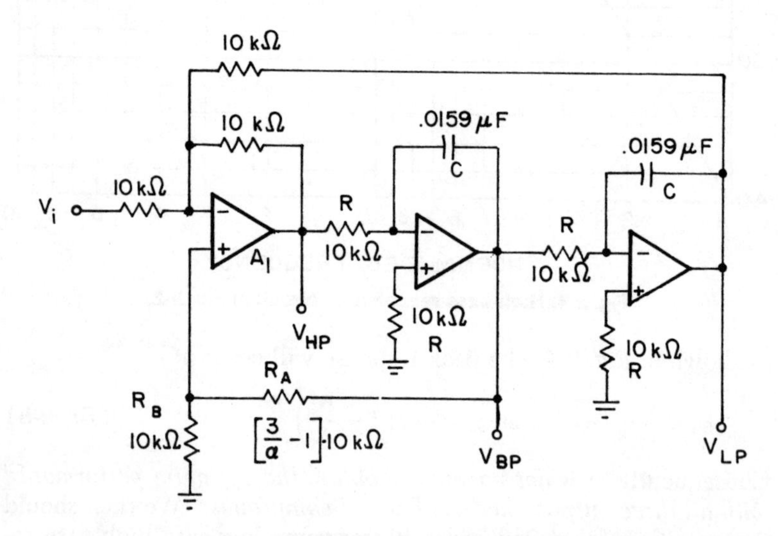

Fig. 9-5. Circuit used for designing unity-gain state-variable filters.

Since the filter's cutoff/center frequency is determined solely by Equation 9-2 for the circuit of Fig. 9-5, the graph shown in Fig. 9-6 is used to rapidly determine the necessary combinations of R and C for a given frequency.

Example:

Design a unity-gain state-variable filter with a cutoff frequency of 700 Hz with a 2nd order Butterworth response. Repeat this example so that the bandpass response has a Q of 50.

From Fig. 9-6, we see that the best combination for R and C is 6.8 kΩ and C = .033 μF. For a 2nd order Butterworth response, the

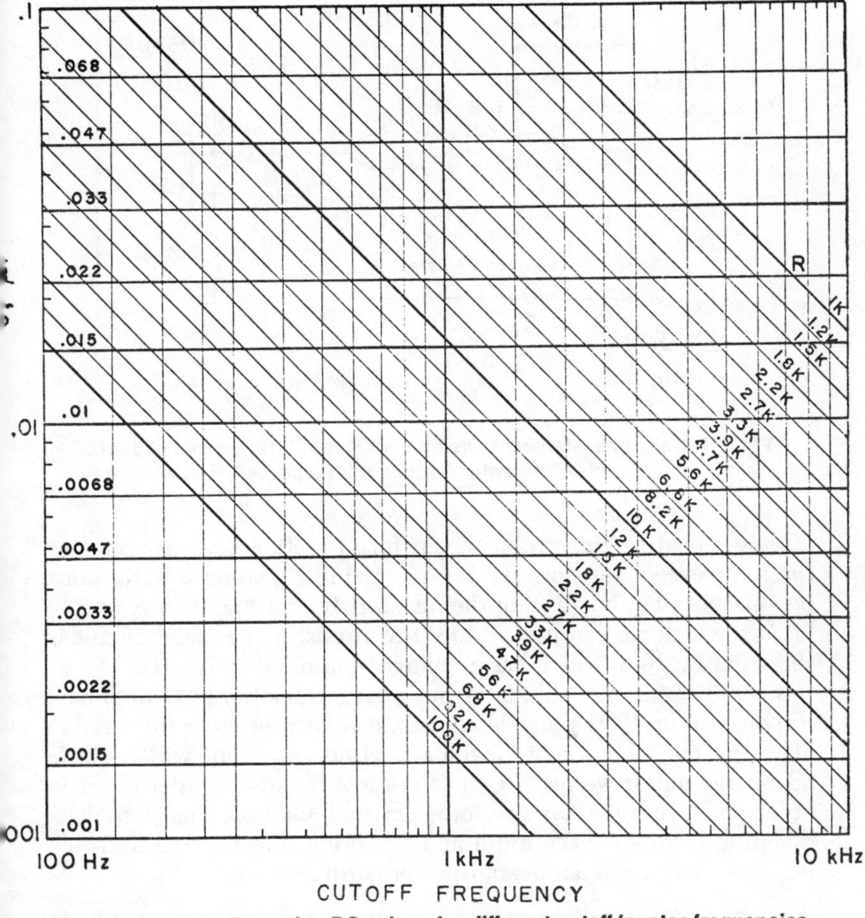

Fig. 9-6. Graph to determine RC values for different cutoff/center frequencies.

damping factor α must be 1.414 so that R_A is $(3/\alpha) - 1$, or 1.12 times R_B.

For minimum offset, R_B is then one-third R, or 2.3 kΩ (a 2.2-kΩ 5% resistor will work). Consequently, R_A is 1.12(2.3 kΩ), or 2.5 kΩ, in which case a 2.4-kΩ 5% resistor will work. If we are willing to neglect the offset, R_B can be any value that is convenient, so long as Equation 9-3 is satisfied. In this case, the combination of $R_A = 3\ k\Omega$ and $R_B = 2.7\ k\Omega$ is a better choice. The final circuit is shown in Fig. 9-7.

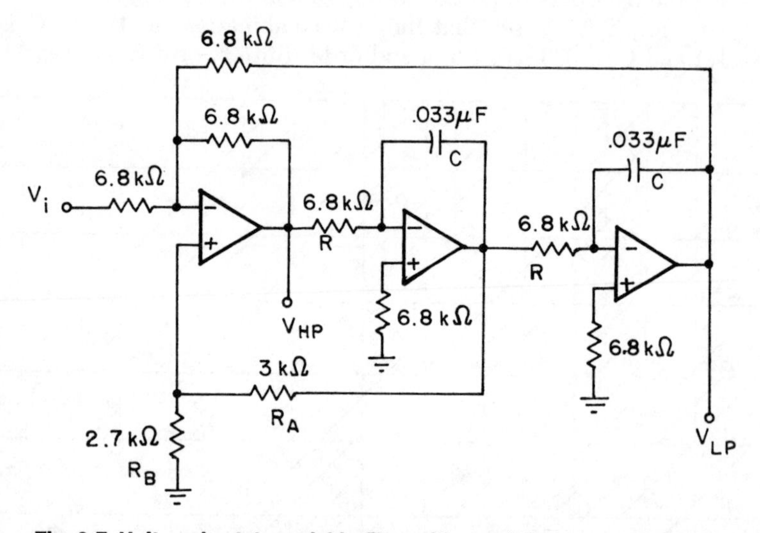

Fig. 9-7. Unity-gain state-variable filter with cutoff frequency of 700 Hz with a 2nd order Butterworth response.

Now repeating this problem, but for $Q = 50$, everything remains the same except resistors R_A and R_B. Using Equation 9-4, R_A must be 149 times R_B. Neglecting the offset, if $R_B = 1\ k\Omega$, then $R_A = 149\ k\Omega$, for which we can use a 150-kΩ 5% resistor. The final circuit is shown in Fig. 9-8, which has a voltage gain of 50 at 700 Hz.

Since the state-variable filter has a 2nd order low-pass/high-pass response, it is then possible to create *higher order* state-variable filters by the methods of Chapter 7. However, there will be little advantage unless we restrict ourselves to 4th and 6th order filters in order to have *simultaneous low-pass and high-pass outputs.* The damping factors for these 4th and 6th order filters, listed in Table 9-1, are used to give an overall Butterworth response.

Table 9-1. Damping Factors for 4th and 6th Order Filters

Order	1st Section	2nd Section	3rd Section
4	1.848	0.765	—
6	1.932	1.414	0.518

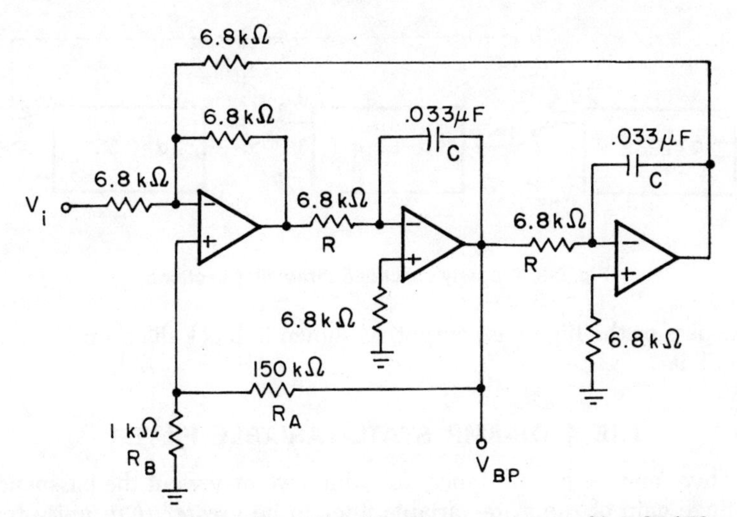

Fig. 9-8. Same filter as Fig. 9-7, but bandpass response with a Q of 50.

Example:

Design a 700-Hz unity-gain 6th order state-variable filter with a Butterworth response.

As with the previous example, R = 6.8 kΩ, and C = .033 μF *for all three sections* (naturally, there will be a total of 9 op amps!). For each section the values of R_A and R_B are found from Equation 9-3:

1st Section: $\alpha = 1.932$

so that $\frac{3}{\alpha} - 1 = 0.55$

or $R_A = 0.55\ R_B$

then $R_A = 1.5\ k\Omega$ and $R_B = 2.7\ k\Omega$ (one possibility)

2nd Section: $\alpha = 1.414$

so that $\frac{3}{\alpha} - 1 = 1.12$

or $R_A = 1.12\ R_B$

then $R_A = 3.6\ k\Omega$ and $R_B = 3.3\ k\Omega$ (again, by trial and error)

3rd Section: $\alpha = 0.518$

so that $\frac{3}{\alpha} - 1 = 4.79$

or $R_A = 4.79\ R_B$

then $R_A = 22\ k\Omega$ and $R_B = 4.7\ k\Omega$

Now to properly cascade these three sections, we need to use a *3-pole-double throw* (*3P2T*) *switch* to be able to select either the

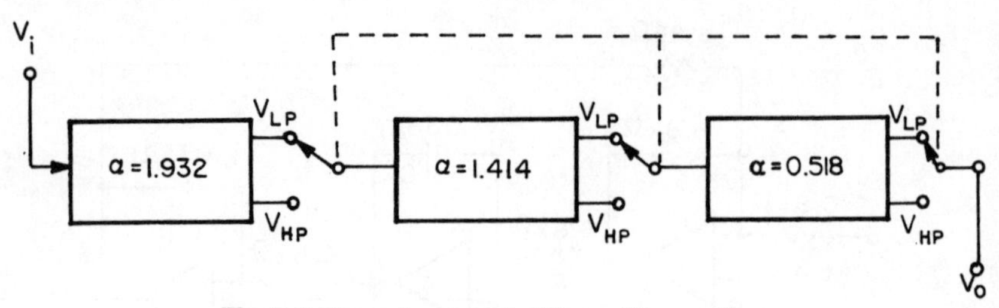

Fig. 9-9. Properly cascaded three filter sections.

low-pass or the high-pass output, as shown in block diagram form in Fig. 9-9.

THE 4 OP-AMP STATE-VARIABLE FILTER

There may be circumstances for which we may want the passband voltage gain of the state-variable filter to be *greater than unity* for the low-pass and high-pass responses. As an initial solution, we can just place an additional op-amp circuit, which can be either an inverting or noninverting amplifier, *after the filter* to give the desired gain.

However, if we change the type of summing amplifier and the damping network of the 3 op-amp state-variable filter, we can then make the gain and damping to be *independent of each other,* as shown in Fig. 9-10 for a cutoff frequency of 1 radian/s and an impedance level of 1Ω. Because of the addition of the 4th op-amp

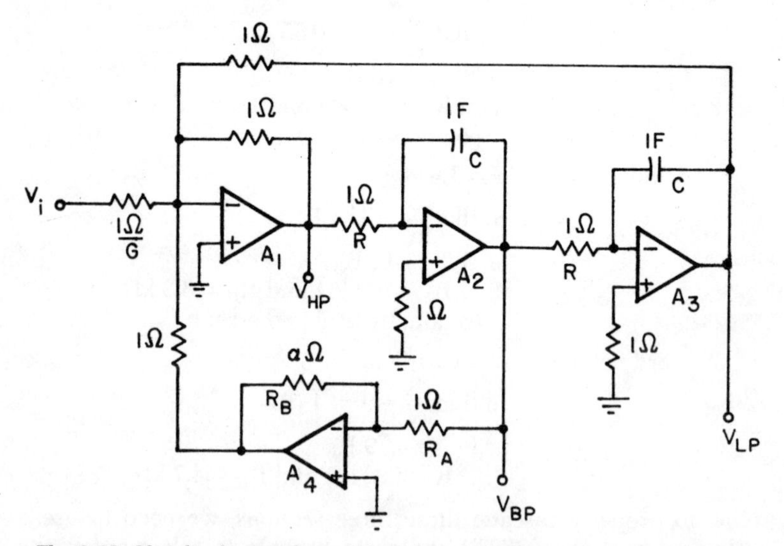

Fig. 9-10. Circuit where gain and damping are independent of each other.

(A_4), the new summing amplifier now has *all three signals connected to its inverting input*, and the $(3/\alpha) - 1$ factor is now eliminated. As before, both integrators (A_2 and A_3) determine the filter's cutoff/center frequency, as given by Equation 9-1.

The output of the summing amplifier A_1 can be expressed in terms of the three inputs as

$$\frac{1}{G}V_{HP} = -V_i - \frac{1}{G}V_{LP} + \frac{1}{G}\left(\frac{R_B}{R_A}\right)V_{BP} \qquad \text{(Eq. 9-14)}$$

or

$$GV_i = -V_{HP} - V_{LP} + \left(\frac{R_B}{R_A}\right)V_{BP} \qquad \text{(Eq. 9-15)}$$

From the above equation, the low-pass and high-pass amplitude responses will have a passband voltage gain equal to

$$\frac{V_{LP}}{V_i} = -G \qquad \text{(Eq. 9-16)}$$

and

$$\frac{V_{HP}}{V_i} = -G \qquad \text{(Eq. 9-17)}$$

respectively. When the last right-hand term of Equations 9-10 and 9-14 are compared, we find that

$$\frac{R_B}{R_A} = \alpha \qquad \text{(Eq. 9-18)}$$

Consequently,

$$R_B = \alpha R_A \qquad \text{(Eq. 9-19)}$$

or

$$R_A = QR_B \qquad \text{(Eq. 9-20)}$$

For the low-pass and high-pass outputs to have a Butterworth response with a passband voltage gain equal to G, then $R_B = 1.414\Omega$ when $R_A = 1\Omega$. For the bandpass filter, the voltage gain at the center frequency will be

$$\frac{V_{BP}}{V_i} = G\left(\frac{R_A}{R_B}\right)$$
$$= GQ \qquad \text{for } \omega = \omega_o \qquad \text{(Eq. 9-21)}$$

and $R_B = Q\Omega$ when $R_A = 1\ \Omega$.

For designing 4 op-amp state-variable filters, we use a circuit that has been normalized to a cutoff frequency of 1 kHz and 10 kΩ, as shown in Fig. 9-11, and illustrated by the following example.

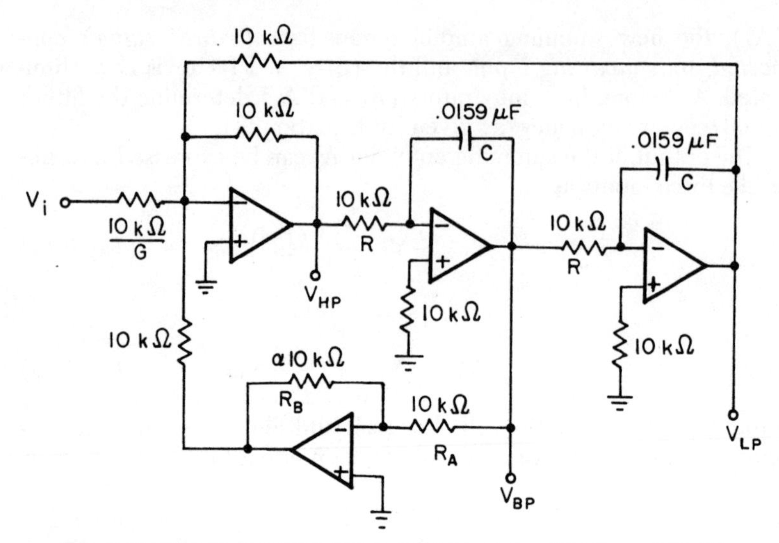

Fig. 9-11. Circuit normalized to cutoff frequency of 1 kHz and 10 kΩ.

Example:

Design a 700-Hz state-variable filter with $Q = 50$, and the voltage gain at this center frequency is to be 100 (+40 dB).

From the previous two examples, we find that $R = 6.8$ kΩ and $C = .033$ μF. Since $Q = 50$ and the voltage gain at the center frequency is 100, then from Equation 9-21,

$$GQ = 100$$

or

$$G = 100/50$$
$$= 2.0$$

Therefore the resistor from the input signal V_i to the inverting input of A_1 must then be equal to 6.8 kΩ/G or 3.4K (we use a 3.3 kΩ 5% resistor).

The values for R_A and R_B are then found from Equation 9-20 by first letting $R_B = 6.8$ kΩ, so that

$$R_A = 50 \times 6.8 \text{ k}\Omega$$
$$= 340 \text{ k}\Omega \qquad \text{(use 330 k}\Omega\text{ 5\% resistor)}$$

and the final circuit is shown in Fig. 9-12.

THE STATE-VARIABLE NOTCH FILTER

One very nice feature of the state-variable filter, whether it be 3 or 4 op amps, is that we can *simultaneously add the low-pass and*

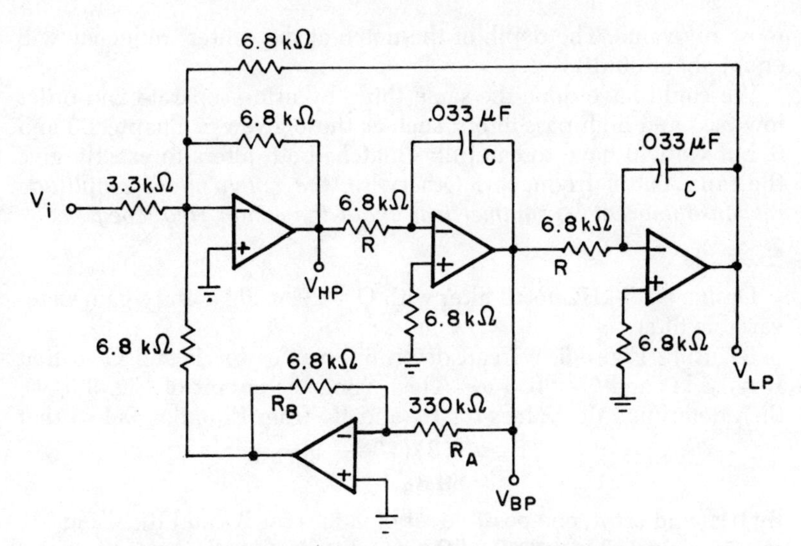

Fig. 9-12. A 700-Hz state-variable filter with a Q of 50.

high-pass outputs equally, obtaining a state-variable notch filter, shown diagrammatically in Fig. 9-13.

What is now needed is a *2-input summing amplifier with equal gains.* Since we already know how to design one from Chapter 2, the final notch filter is shown in Fig. 9-14, and the resistors R can be al-

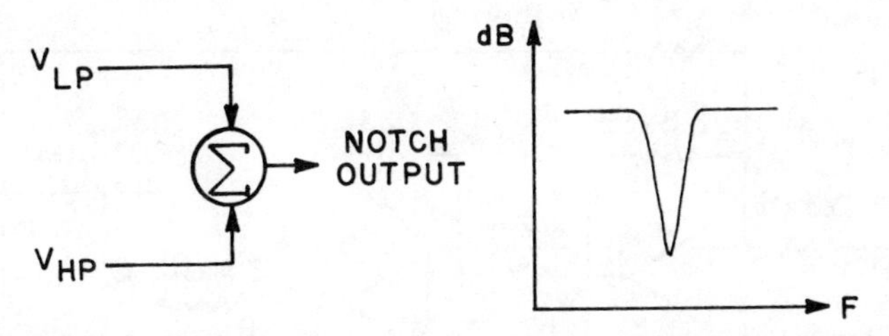

Fig. 9-13. A state-variable notch filter shown diagrammatically.

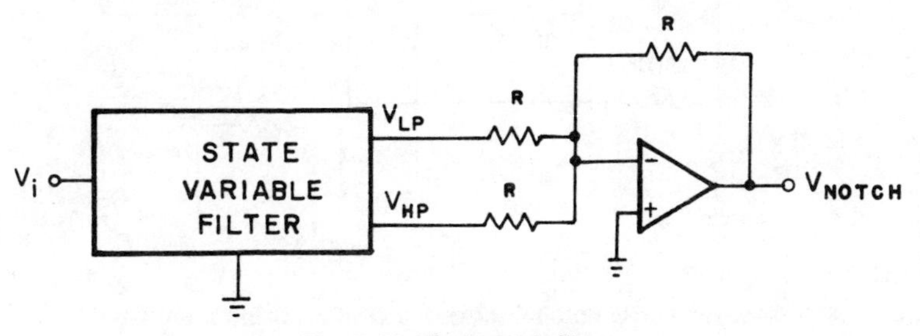

Fig. 9-14. Final notch filter.

most any value. The depth of the notch at the center frequency will equal about 30 dB.

We could have done the same thing by using separate 2nd order low-pass and high-pass filters, such as those given in Chapters 5 and 6, but we will have to carefully "match" both filters to exactly give the same cutoff frequency. Otherwise, *the notch filter's amplitude response will not be symmetrical about the center frequency.*

Example:

Design a 1.5-kHz notch filter with Q = 20, using a unity-gain state-variable filter.

By using Fig. 9-6, we can determine values for R and C so that R = 3.3 kΩ and C = .033 μF. Then using the circuit of Fig. 9-5, we then determine the values for R_A and R_B from Equation 9-4 so that

$$R_A = [(3)(20) - 1]R_B$$
$$= 59\ R_B$$

By trial and error, one possible set of values for R_A and R_B, using 5% resistors, are 130 kΩ and 2.2 kΩ respectively. The design of the state-variable section is done. To complete the design, add the 2-input op-amp summing amplifier to the low-pass and high-pass outputs, giving the final circuit of Fig. 9-15.

Example:

Repeat the previous example, but with the 4-op-amp state-variable filter, so that the voltage gain in the passband is unity (0 dB).

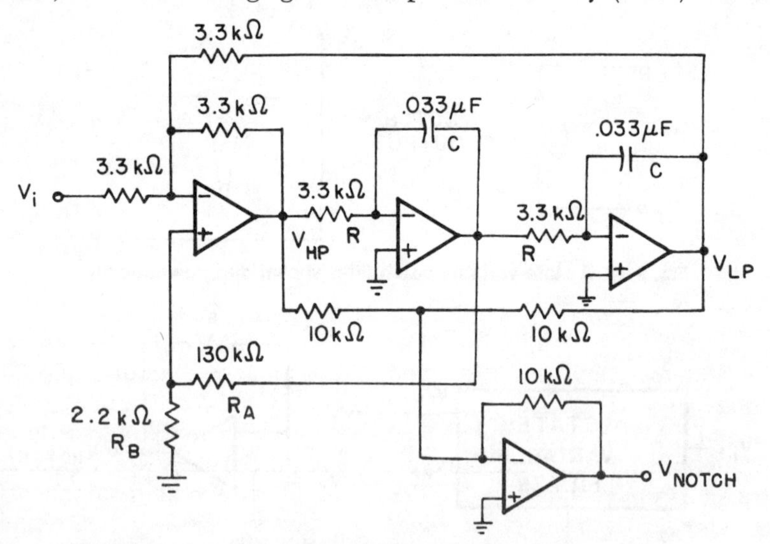

Fig. 9-15. A 1.5-kHz notch filter with a Q of 20, using a unity-gain state-variable filter.

Using Fig. 9-11, the values for R and C are the same as for the previous example. Since the passband gain is 1, the input resistor, i.e. 3.3 kΩ/G, is 3.3 kΩ. With Q = 20, Equation 9-20 then determines the values for R_A and R_B,

$$\begin{aligned} R_A &= QR_B \\ &= (20)(3.3\text{ k}\Omega) \\ &= 66\text{ k}\Omega \qquad (\text{use } 68\text{ k}\Omega, 5\%) \end{aligned}$$

As before, all that is needed now is a 2-input summing amplifier, giving the final circuit of Fig. 9-16.

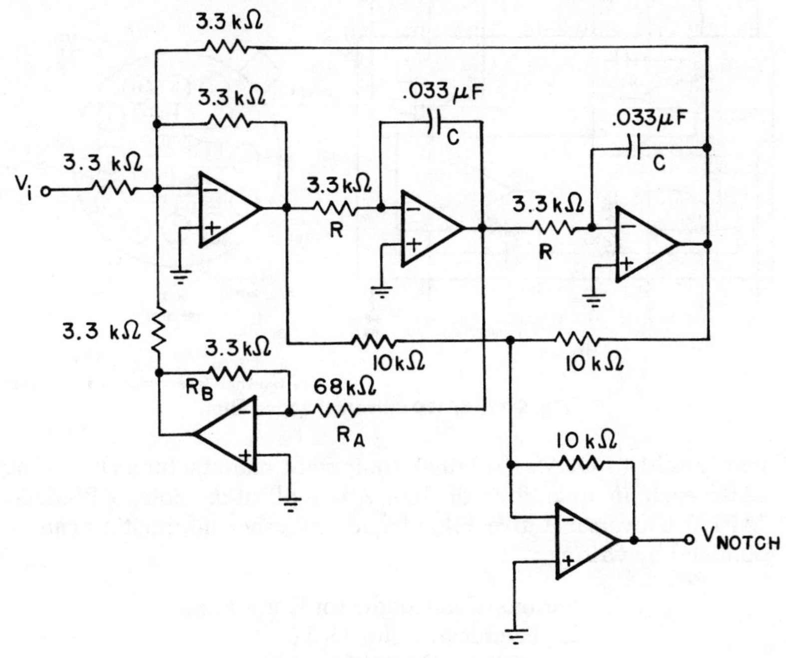

Fig. 9-16. A 4 op-amp notch filter.

COMMERCIAL FILTERS

With the advancement of solid state technology, a number of manufacturers now offer "ready-to-go" state-variable filters in either metal cans, or DIP units. Depending on whether the unity-gain or the 4 op-amp state-variable types are required, only 3 or 4 external resistors are required to "program" the filter to your requirements.

The commercial devices essentially follow the design that is described in this chapter, except that the frequency-determining capacitors and the resistors associated with the summing amplifier are already inside the filter. An additional op-amp section, which is uncommitted, can be used to form notch filters. Some of the commercial types available are listed below.

National Semiconductor Corporation

The model AF100 Universal Active Filter, made by National Semiconductor, comes in either an 11-pin type TO-8 can, or a 16-pin DIP integrated circuit, as illustrated below in Fig. 9-17. This unit is primarily sold to OEMs (original equipment manufacturers), costing $4.95 each in quantities of 100. An application note, #B5M26—"AF100 Universal Active Filter," and any other information can be obtained by writing:

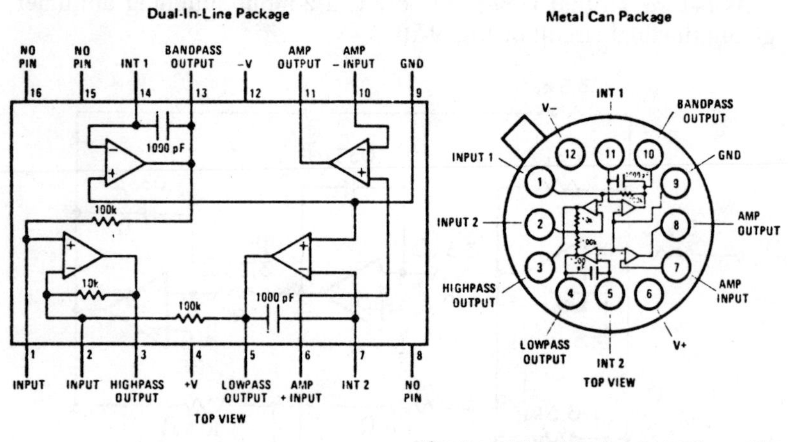

Courtesy National Semiconductor Corporation

Fig. 9-17. AF100 universal active filter.

National Semiconductor Corporation
2900 Semiconductor Drive
Santa Clara, CA 95051

Burr-Brown

The model UAF41 state-variable filter, made by Burr-Brown, is similar to the AF100, except that it comes in a 14-pin DIP package, as shown in Fig. 9-18. In quantities of 100, the UAF41 sells for $8.00 each. An applications note, #PDS-359—"UAF41 Universal Active Filter," as well as other information, can be obtained from:

Burr-Brown
P.O. Box 11400
Tucson, AZ 85734

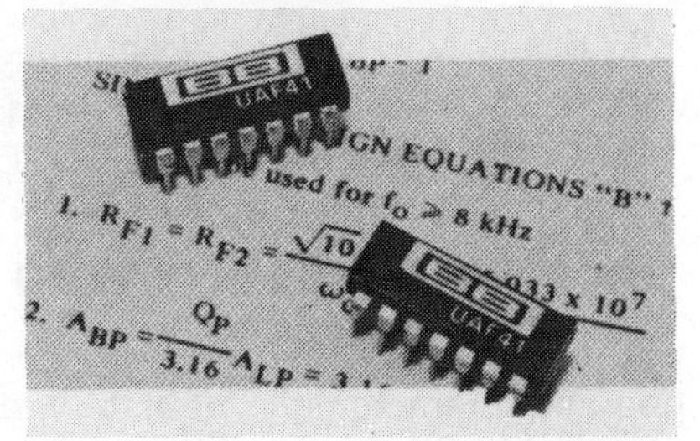

Fig. 9-18. UAF41 state-variable filter.

Courtesy Burr-Brown

General Instrument Corporation

General Instrument manufactures several state-variable filters:

ACF 7092, 16-pin DIP
ACF 7096, 14-pin DIP
ACF 7032, 16-pin TO-8 can

and all are similar to the types made by National and Burr-Brown. A recent article covering its design is:

Volpe, G. T., and L. Premisler. "Universal Building Blocks Simplify Active Filter Design," *EDN*, September 5, 1976, pp. 91-95.

or you can request the company's applications note, #HY1001—"Series 7000 Active Filters," by writing:

General Instrument Corporation
Semiconductor Products Group
Hybrid Microcircuit Department
600 W. John Street
Hicksville, NY 11802

Kinetic Technology Incorporated

The model FS-60 Active Filter, made by KTI, is a 14-pin DIP device and sells for $32.00. However, a cull version, the model FX-60, is available for $6.00 each from:

Compar Chesapeake, Inc.
7 Church Lane
Baltimore, MD 21208

Two articles that briefly outline the design rules are:

Holladay, K. "Tunable Audio Filter for Weak-Signal Communications," *Ham Radio*, November, 1975, pp. 28-34.

Chapman, M. A. "Audio Filters for Improving SSB and CW Reception," *Ham Radio*, November, 1976, pp. 18-23.

However, design information can be obtained from:

Kinetic Technology, Incorporated
3393 De La Cruz Boulevard
Santa Clara, CA 95050

AN INTRODUCTION TO THE EXPERIMENTS

The following experiments are designed to demonstrate the operation and characteristics of state-variable filters. As with the experiments of the previous chapter, the numerical calculations required will be greatly simplified by the use of a pocket calculator.

The experiments that you will perform can be summarized as follows:

Experiment No.	*Purpose*
1	Demonstrates the operation and characteristics of the unity-gain state-variable filter.
2	Demonstrates the operation and characteristics of the 4 op-amp state-variable filter.
3	Demonstrates the operation and characteristics of a state-variable notch filter.

EXPERIMENT NO. 1

Purpose

The purpose of this experiment is to demonstrate the operation and characteristics of a unity-gain state-variable filter.

Schematic Diagram of Circuit (Fig. 9-19)

Design Basics

- Cutoff frequency: $f_c = 1/2\pi RC$
- $R_A = [3Q - 1]R_B$, where $Q = 1/\alpha$
- Passband gains:
    - Lowpass $= -1$ (180° out of phase)
    - Highpass $= -1$
    - Bandpass $= Q$ (in phase)

Step 1

Set your oscilloscope for the following initial settings:

- Channels 1 & 2: 0.2 V/division
- Time base: 1.0 ms/division
- AC coupling

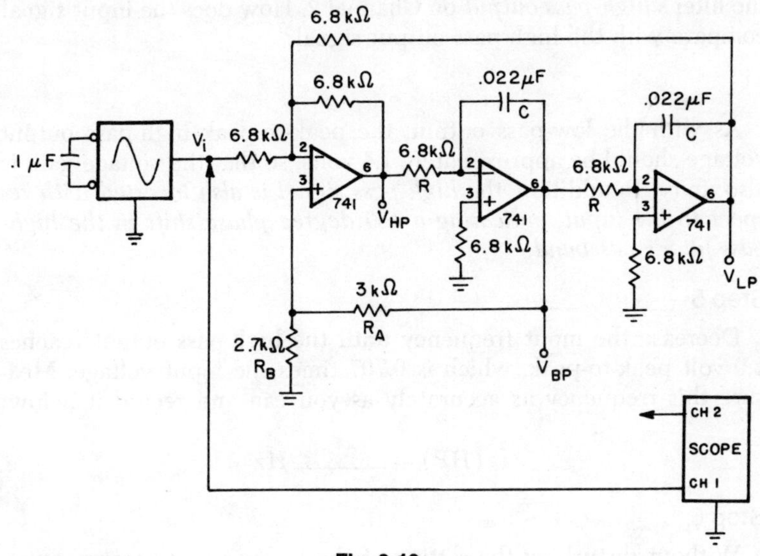

Fig. 9-19.

Step 2

Wire the circuit shown in the schematic diagram. If you don't have a 3-kΩ resistor, use two 1.5-kΩ resistors *connected in series.*

Step 3

Apply power to the breadboard and first adjust the input signal at 1.4 volts peak-to-peak (7 vertical divisions)—*make this setting as accurate as you can!* Then set the input frequency at 100 Hz (1 comlete cycle/10 horizontal divisions). Next measure the *low-pass output* on Channel 2 of the oscilloscope. How does the input signal compare with the low-pass output signal?

You should find that the output signal's amplitude is essentially the same as the input's, or 1.4 volts. Consequently, the voltage gain for the low-pass output is *unity.* In addition, *the low-pass output signal is inverted with respect to the input, so that the two signals are out of phase by 180 degrees in the passband.*

Step 4

Now set the oscilloscope's time base at 0.1 ms/division and adjust the input frequency so that 1 complete cycle occupies the 10 horizontal divisions (a frequency of 10 kHz). Also check to make sure that the input voltage is still 1.4 volts peak-to-peak. Then measure the filter's *high-pass output* on Channel 2. How does the input signal compare with the high-pass output signal?

As with the low-pass output, the peak-to-peak high-pass output voltage should be approximately 1.4 volts, so that the voltage gain is also unity. In addition, the *high-pass signal is also inverted with respect to the input, indicating a 180 degree phase shift in the high-pass filter's passband.*

Step 5

Decrease the input frequency until the high-pass output reaches 1.0 volt peak-to-peak, which is 0.707 times the input voltage. Measure this frequency as accurately as you can and record it below:

$$f_c\ (HP) = \_\_\_\_\_\_ \text{ Hz}$$

Step 6

Without disturbing the setting of the frequency generator, transfer the probe connected to the high-pass output to the low-pass output of the filter and measure the voltage. Is it 1.0 volt? If not, vary the generator's frequency slightly so that the output is 1.0 volt peak-to-peak. Measure this frequency as accurately as you can and record it below:

$$f_c\ (LP) = \_\_\_\_\_\_ \text{ Hz}$$

If the frequency-determining components of both integrators are fairly well matched, the frequencies measured in Steps 5 and 6 should almost be the same, to within several Hz. How do these two frequencies compare with the expected frequency?

Depending on the accuracy of your measurements, or the quality of your components, you should have measured frequencies near the theoretical value of 1064 Hz. When I performed this experiment, my measured values were 971 Hz (for Step 5) and 968 Hz (for Step 6). Although they were "a little bit" off, the two integrator sections were fairly well matched, since there was a difference of only 3 Hz between the two measured values.

Step 7

Now transfer the scope's Channel 2 probe to the filter's bandpass output. Very carefully vary the generator frequency up and down, stopping at the point at which the output voltage is a maximum. Now measure this frequency as carefully as you can and record it below:

$$f_o\ (BP) = \_\_\_\_\_\_\ Hz$$

If the two measured frequencies of Steps 5 and 6 are different, indicating the two integrators are mismatched to some degree, we must then find the filter's center frequency by taking the *geometric average* of these two frequencies,

$$f_o = (f_{HP} f_{LP})^{1/2}$$

The geometric average of the two frequencies that I measured was found to be 969.5 Hz, while the measured value for f_o was 970 Hz—*Not too bad!* If you found that the frequencies were equal, in Steps 5 and 6, how does the value that you determined in this step compare?

Within a few hertz, they should almost be the same.

Step 8

Without disturbing anything, measure the filter's bandpass output voltage. What is the voltage gain at the bandpass filter's center frequency? Is it what you would expect?

You should have measured an output voltage of approximately 1.0 volt, so that the voltage gain will be 0.70 at the center frequency, and numerically should be equal to the filter's Q, or, taking its reciprocal, equal to the damping factor α, or very nearly 1.41. Consequently, we should conclude that this state-variable filter is primarily designed to be used as either a low-pass or high-pass Butterworth filter. From the equation in the "design basics" section relating resistors R_A and R_B, we find that Q should equal 0.704 with the component values shown. In addition, you should observe that the bandpass output signal, at the center frequency, *is in phase with the input signal.*

Step 9

Disconnect the power to the breadboard. Now replace the 3-kΩ resistor (R_A) with a 27-kΩ resistor and apply the power to the breadboard. Repeat Steps 1, 3, 4, 5, 6, 7, and 8 of this experiment. How do your results compare with the design equations?

You should find no change in the center frequency, the low-pass and high-pass cutoff frequencies, or the passband gain for either the low-pass or high-pass outputs. However the bandpass filter's Q has increased to approximately 3.7, and is also the bandpass filter's gain at the center frequency.

Step 10

If time permits, try another value for R_A, for example 270 kΩ, and see if the design equations still hold.

EXPERIMENT NO. 2

Purpose

The purpose of this experiment is to demonstrate the operation and characteristics of the 4 op-amp state-variable filter.

Schematic Diagram of Circuit (Fig. 9-20)

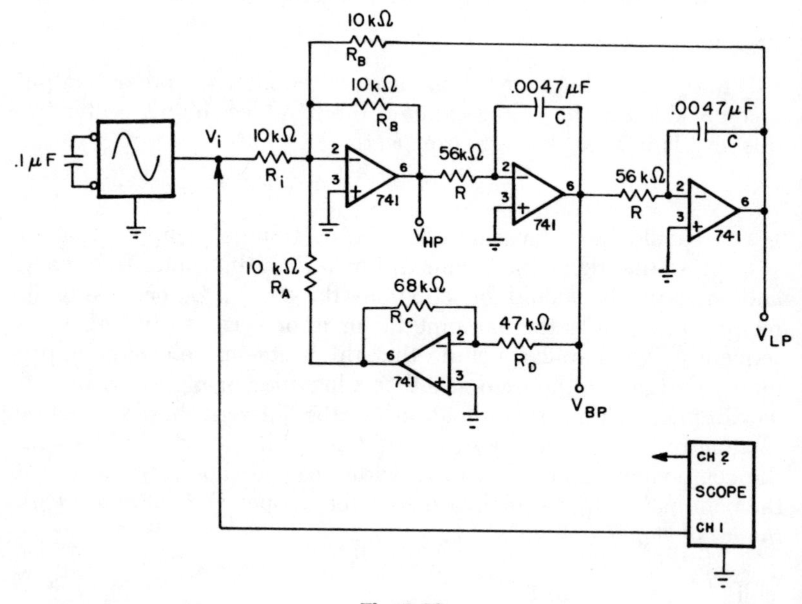

Fig. 9-20.

Design Basics

- Cutoff and center frequency: $f_c = 1/2\pi RC$
- $R_D = Q\ R_C$ or $R_C = \alpha\ R_D$ (since $Q = 1/\alpha$)
- Passband gains: set by the ratio $R_B/R_i = G$
    - Low-pass $= -G$ (180° out of phase)
    - High-pass $= -G$
    - Bandpass $= GQ$ (in phase)

Step 1

Set your oscilloscope for the following initial settings:

- Channels 1 & 2: 0.1 V/division
- Time base: 1 ms/division
- AC coupling

Step 2

Wire the circuit shown in the schematic diagram. Apply power to the breadboard and adjust the input signal at 0.7 volt peak-to-peak (7 vertical divisions). Then set the input frequency at 100 Hz (1 complete cycle/10 horizontal divisions).

Step 3

Now measure the output voltage of the *low-pass section* (at pin 6 of the last op amp) with the probe connected to Channel 2 of your oscilloscope. How does the input signal compare (amplitude and phase) with the low-pass output signal?

The low-pass output voltage should be very nearly, if not exactly equal to the input voltage, or 0.7 volt peak-to-peak. Since, from the design equations, the cutoff frequency of the low-pass section is approximately 600 Hz, the frequency of 100 Hz is within the low-pass filter's *passband,* and your measured passband gain should be very nearly 1.0. In addition, there is a 180° phase shift, since the output voltage is inverted with respect to the filter's input.

For the passband voltage gain to be unity, *all of the input resistors of the summing amplifier must be equal to the feedback resistor,* R_B, *so that* $R_i = R_A = R_B$. In a later step, we shall see that the passband gain is determined by the relationship:

$$G = R_B/R_i$$

where R_B is always equal to R_A.

Step 4

Now connect the "Channel 2" probe to the filter's *bandpass output,* and set the oscilloscope time base at 0.2 ms/division. Carefully adjust the generator frequency until the bandpass output (Channel 2) reaches its maximum value. Determine the voltage gain of the bandpass section and record it below:

$$G_{(BP)} = \_\_\_\_\_\_$$

You should have measured a peak-to-peak voltage of approximately 0.5 volt, so that the passband voltage gain of the bandpass section is 0.5/0.7, or 0.71. From the "design basics" section, the filter's Q is determined solely by the ratio of resistors R_C and R_D. In terms of the values used presently in this experiment,

$$Q = \frac{R_D}{R_C} = \frac{47\ k\Omega}{68\ k\Omega} = 0.69$$

and since the resistance ratio R_B/R_i is 1.0, the passband gain of the bandpass section is also equal to the filter's Q, or 0.69. In addition, since $\alpha = 1/Q$, the damping factor for the low-pass and high-pass sections is equal to 1.447 (using the values for R_C and R_D in this experiment). Consequently, the low-pass and high-pass responses should almost equal that of a 2nd order Butterworth filter. Using standard 5% resistors, this is the best combination that will give a 2nd order Butterworth response, being less than 3% from the ideal value of 1.414.

Step 5

The generator frequency is now set at the filter's *center frequency*, since the output voltage of the bandpass section is at its maximum value. Determine this center frequency and record it below:

$$f_o = \_\_\_\_\_\_\_\_ \text{Hz}$$

You should have measured a value near the theoretical frequency of 605 Hz. In addition, you should have also noted that the output signal is *in phase* with the filter's input signal. Is it?

Step 6

Change the scope time base to 20 μs/division, and adjust the input signal so that 1 cycle occupies 10 horizontal divisions, or a frequency of 5000 Hz. Now connect the "Channel 2" probe to the filter's *high-pass output*. What is the voltage gain? How does it compare with the value you found for the low-pass output in Step 3?

At 5000 Hz, which is almost *1 decade above the high-pass filter's cutoff frequency of approximately 605 Hz, you should have obtained the same results as in Step 3.*

Step 7

Now change the resistor marked "R_i" in the schematic diagram (10 kΩ) to 1 kΩ. All other 10-kΩ resistors are left the same! In addition, change the setting of Channel 2 to 1 V/division, and the time base back to 1 ms/division. Finally, adjust the input voltage at 0.4 volt and at a frequency of 100 Hz. Measure the output voltage of the *low-pass section,* and determine the voltage gain, recording your value below:

$$G_{(LP)} = \_\_\_\_\_\_$$

You should have calculated a voltage gain of approximately 10.0. Why?

Since the 1-kΩ resistor is 1/10 the value of the other 10-kΩ input resistors, as well as the 10-kΩ feedback resistor of the summing amplifier, the gain is then 10. If, for example, resistor "R_i" were 2.7 kΩ, the gain would be, according to the formula:

$$G = \frac{10\ k\Omega}{2.7\ k\Omega} = \frac{R_B}{R_i}$$

$$= 3.7$$

Consequently, we can change the passband gain of the filter simply by changing the value of R_i.

Step 8

Now connect the "Channel 2" probe to the filter's *bandpass output,* and change the setting of Channel 2 to 0.5 V/division. As in Step 4, carefully adjust the generator frequency until the bandpass output voltage reaches a maximum value. Determine the voltage gain for this section and record it below:

$$G_{(BP)} = \_\_\_\_\_\_$$

How does it compare with the value you determined in Step 4?

The gain that you have just determined should be approximately 10 times that of Step 4. When this experiment was performed, I determined a gain of 0.70 in Step 4, and a gain of 7.00 for this step. If we now divide the above value by the passband gain of the low-pass response (determined in Step 7), which should be approximately 10, we now have the Q of the bandpass response, or

$$Q = \frac{G_{(BP)}}{G_{(LP)}}$$

By decreasing R_i by a factor of 10, *all we have done is to increase the passband gain of the low-pass, high-pass, and bandpass responses by 10. We have in no way changed the filter's Q or damping factor, which is set solely by resistors R_C and R_D.*

Step 9

Disconnect the power to the breadboard. Now remove the 1-kΩ resistor (R_i) and, in its place, connect two 10-kΩ resistors *in parallel*, so that the equivalent resistance for R_i is now 5 kΩ. Also, replace the 47 kΩ resistor (R_D) with a 680 kΩ resistor. Before we go any further, let's analyze what has just been done to our circuit.

Since R_i is now 5 kΩ, the passband gain G is now expected to be 2.0, since

$$G = \frac{R_B}{R_i} = \frac{10\ \text{k}\Omega}{5\ \text{k}\Omega} = 2.0$$

Also, since resistor R_D is now 680 kΩ, and R_C is still 68 kΩ, the Q is now

$$Q = \frac{R_D}{R_C} = \frac{680\ \text{k}\Omega}{68\ \text{k}\Omega} = 10$$

Consequently, the output voltage of the bandpass section will be GQ or 20 times the input voltage.

Step 10

Now set your oscilloscope for the following settings:

- Channel 1: 20 mV/division
- Channel 2: 50 mV/division
- Time base: 1 ms/division

Step 11

Apply power to the breadboard and very carefully adjust the input voltage at 0.14 volt (7 vertical divisions) and the frequency at 100 Hz (1 cycle/10 horizontal divisions). Connect the "Channel 2" probe to the filter's *low-pass output*. Determine the voltage gain and record it below:

$$G_{(LP)} = \_\_\_\_\_\_$$

As was discussed in Step 9, the gain should be approximately 2.0, since R_i is now 1/2 R_B.

Step 12

Now connect the "Channel 2" probe to the filter's *bandpass output.* Set Channel 2 at 0.5 V/division and the time base at 0.2 ms/division. Very carefully vary the generator frequency until the output voltage of the bandpass section reaches its maximum output, and measure the voltage gain, recording its value below:

$$G_{(BP)} = \_\_\_\_\_\_$$

You should have measured an output voltage of approximately 2.8 volts (i.e, 8 vertical divisions) so that the voltage gain is then 2.8/.14, or 20. When I performed this experiment, I measured a gain of 19.3. Now divide your measured value by the gain you determined in Step 11, and record it below:

$$G_{(BP)}/G_{(LP)} = \_\_\_\_\_\_$$

This ratio is then the filter's Q, which should be approximately 10, as was analyzed in Step 9.

Step 13 (Optional)

If time permits, experimentally verify the filter's Q by measuring the center frequency and the 3-dB frequencies. Your value should be approximately the same as the value determined in Step 12.

Step 14

Save this circuit, as it will be used in the next experiment with some extra components.

EXPERIMENT NO. 3

Purpose

The purpose of this experiment is to demonstrate the design and operation of a state-variable notch filter, formed by summing the low-pass and high-pass outputs of a 4 op-amp state-variable filter.

Schematic Diagram of Circuit (Fig. 9-21)

Design Basics

- Cutoff frequency: $f_c = 1/2\pi RC$
- $R_D = Q\ R_C$

Step 1

Set your oscilloscope for the following initial settings:

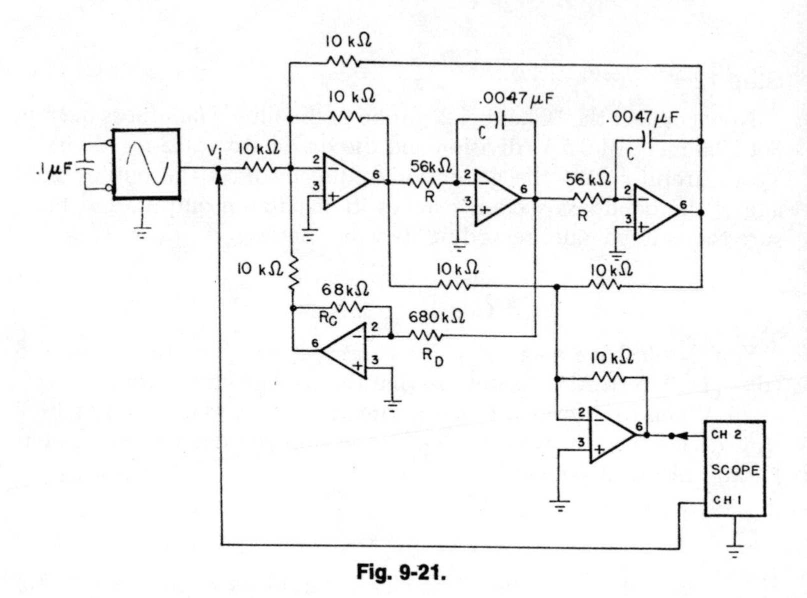

Fig. 9-21.

- Channel 1: 0.2 V/division
- Channel 2: 20 mV/division
- Time base: 1 ms/division
- AC coupling

Step 2

Wire the circuit shown in the schematic diagram. Apply power to the breadboard and set the input signal at 1.4 volts peak-to-peak (7 vertical divisions).

Step 3

Slowly vary the generator frequency until the output voltage of the summing amplifier reaches a minimum value, as seen on Channel 2 of the oscilloscope. Now measure the output voltage and determine the depth of null, recording your result below:

depth of null = ______ dB

As a comparison, I experienced a null depth of −26.7 dB.

Step 4

Now set the oscilloscope time base to 0.2 ms/division and determine the null frequency, recording your result below:

f_n = ______ Hz

How does this null frequency compare with the value found in Step 5 of the previous experiment?

The null frequency of this experiment and the center frequency of the previous experiment should be approximately the same. This is because the state-variable filter section is the same for both experiments, and the frequency determining equations are also equal.

Step 5

Change the setting of Channel 2 to 0.2 V/division. Since the passband gain of this filter has been set at unity, for convenience, vary the generator frequency above and below the filter's null frequency so that the output voltage is 0.707 times the input voltage, or 1.0 volt (or −3 dB), and record your results below:

$$f_L = \_\_\_\_\_\_ \text{ Hz}$$

$$f_H = \_\_\_\_\_\_ \text{ Hz}$$

As a simple check, determine the null frequency based on the above two values according to the equation:

$$f_n = (f_L f_H)^{1/2}$$

Within 5%, does this calculated value agree with the experimental value of Step 4?

Step 6

Using the values for f_L and f_H of Step 5, and f_n of Step 4, determine the notch filter's Q from the relation

$$Q = \frac{f_n}{f_H - f_L}$$

and record your result below:

$$Q = \_\_\_\_\_\_$$

According to the equation given in the "design basics" section, you should have determined a Q of 10, within 5%. This is the same value determined in Step 9 of the previous experiment.

It should be pointed out that a state-variable notch filter can also be made by connecting the same 2-input summing amplifier to the low-pass and high-pass outputs of the *unity-gain* state-variable filter, which is frequently the case. We have used the 4 op-amp state-variable filter for convenience, since we had been using it in the previous experiment.

With this state-variable notch filter, we now have simultaneously the four major types of filter responses discussed in this book: low-pass, high-pass, bandpass, and notch.

Suggested References

The following books and short articles contain additional material on the design and use of active filters. Although not a complete list, these references for the most part are devoid of heavy mathematical theory.

BOOKS

1. Jung, W. G. *IC Op-Amp Cookbook.* Indianapolis: Howard W. Sams & Co., Inc., 1974.
2. Lancaster, D. *Active-Filter Cookbook.* Indianapolis: Howard W. Sams & Co., Inc., 1975.
3. Lenk, J. D. *Manual for Operational Amplifier Users.* Reston: Reston Publishing Co., 1976.
4. Tobey, G. E., Graeme, J. G., and L. P. Huelsman. *Operational Amplifiers—Design and Applications.* New York: McGraw Hill, 1971.

SHORT ARTICLES

1. Bainter, J. R. "Active filter has stable notch and response can be regulated." *Electronics,* October 2, 1975, pp. 115-117.
2. Chapman, M. A. "Audio filters for improving SSB and CW reception." *Ham Radio,* November, 1976, pp. 18-23.
3. Cushing, P. "Tunable active filter has switchable response." *Electronics,* January 4, 1973, p. 104.
4. Derrett, C. J., and P. J. Tavner. "Active filters: some simple design methods for biomedical workers." *Medical and Biological Engineering,* November, 1975, pp. 883-888; also May, 1976 (errata), p. 364.
5. Griffee, F. M. "RC active filters using op-amps." *Ham Radio,* October, 1976, pp. 54-58.
6. Hall, C. "Tunable RC notch filter." *Ham Radio,* September, 1975, pp. 16-20.
7. Holladay, K. "Tunable audio filter for weak-signal communications." *Ham Radio,* November, 1975, pp. 28-34.

8. Kesner, D. "An introduction to active filters." *CQ,* April, 1975, pp. 32-34, and 66-68.
9. Kraus, K. "Bandpass/lowpass filter with two op amps." *Electronic Engineering,* November, 1975, p. 25.
10. Lancaster, D. "Understanding active filters." *Popular Electronics,* December, 1976, pp. 69-73.
11. Lefferson, P. "Tunable notch filter suppresses hum." *Electronics,* September 2, 1976, p. 100.
12. Moberg, G. O. "Multiple-feedback filter has low Q and high gain." *Electronics,* December 9, 1976, pp. 97-99.
13. Nicosia, N. J. "A tunable audio filter for CW." *Ham Radio,* August, 1970, pp. 34-35.
14. Russell, H. T. "Design active filters with less effort." *Electronic Design,* January 7, 1971, pp. 82-85.
15. Schaeffer, L. "Op-amp active filters—simple to design once you know the game." *EDN,* April 20, 1976, pp. 79-84.
16. Stremler, F. G. "Simple arithmetic: an easy way to design active bandpass filters." *Electronics,* June 7, 1971, pp. 86-89.
17. Volpe, G. T., and L. Premisler. "Universal building blocks simplify active filter design," *EDN,* September 5, 1976, pp. 91-95.

APPENDIX A

Standard Resistor and Capacitor Values

RESISTORS

1. The following ±5% standard decade values are available. Those marked with * are most commonly obtained from electronic suppliers.

1.0*	1.8*	3.3*	5.6*
1.1	2.0	3.6	6.2
1.2*	2.2*	3.9*	6.8*
1.3	2.4	4.3	7.5
1.5*	2.7*	4.7*	8.2*
1.6	3.0	5.1	9.1

To obtain standard resistance values, multiply preferred number from decade table by powers of 10. Standard values are available from 10Ω to 22 MΩ.

2. The following ±1% values are available, but at a higher cost.

10.0	12.1	14.7	17.8	21.5	26.1	31.6	38.3	46.4	56.2	68.1	82.5
10.2	12.4	15.0	18.2	22.1	26.7	32.4	39.2	47.5	57.6	69.8	84.5
10.5	12.7	15.4	18.7	22.6	27.4	33.2	40.2	48.7	59.0	71.5	86.6
10.7	13.0	15.8	19.1	23.2	28.0	34.0	41.2	49.9	60.4	73.2	88.7
11.0	13.3	16.2	19.6	23.7	28.7	34.8	42.2	51.1	61.9	75.0	90.9
11.3	13.7	16.5	20.0	24.3	29.4	35.7	43.2	52.3	63.4	76.8	93.1
11.5	14.0	16.9	20.5	24.9	30.1	36.5	44.2	53.6	64.9	78.7	95.3
11.8	14.3	17.4	21.0	25.5	30.9	37.4	45.3	54.9	66.5	80.6	97.6

These standard values are available from 10Ω to 22.1 MΩ.

CAPACITORS

Capacitor values follow the standard 10% decade values for resistors. For active filters, the capacitors should be either Mylar or Tantalum. Those values marked with * below are most commonly available from electronic suppliers.

.001 μF*	.01 μF*	.1 μF*
.0012 μF	.012 μF	.12 μF
.0015 μF	.015 μF	.15 μF
.0018 μF	.018 μF	.18 μF
.002 μF	.02 μF	.2 μF
.0022 μF*	.022 μF*	.22 μF*
.0025 μF	.025 μF	.25 μF
.0027 μF	.027 μF	.27 μF
.0033 μF*	.033 μF*	.33 μF*
.0039 μF	.039 μF	.39 μF
.0047 μF*	.047 μF*	.47 μF*
.005 μF	.05 μF	.5 μF
.0056 μF	.056 μF	.56 μF
.0068 μF*	.068 μF*	.68 μF*
.0075 μF	.075 μF	.75 μF
.0082 μF	.082 μF	.82 μF

APPENDIX B

Derivations

A. 1ST ORDER FILTER (FIG. B-1)

General Formula:

At node A, using a voltage divider relationship,

$$V_A = \frac{1/Y_2}{1/Y_1 + 1/Y_2} V_i \qquad \text{(Eq. A-1)}$$

$$= \frac{Y_1}{Y_1 + Y_2} V_i$$

Also,

$$V_o = GV_A \qquad \text{(Eq. A-2)}$$

where G is the passband gain (the closed loop, or noninverting gain), so that

$$V_o = \frac{GY_1}{Y_1 + Y_2} V_i$$

or

$$\frac{V_o}{V_i} = \frac{GY_1}{Y_1 + Y_2} \qquad \text{(Eq. A-3)}$$

For a Low-Pass Filter

The transfer function describing a 1st order low-pass filter with a passband gain G is given by

$$\frac{V_o}{V_i} = \frac{G\omega}{s + \omega_c} \qquad \text{(Eq. A-4)}$$

Comparing the terms of equations A-3 and A-4 implies that

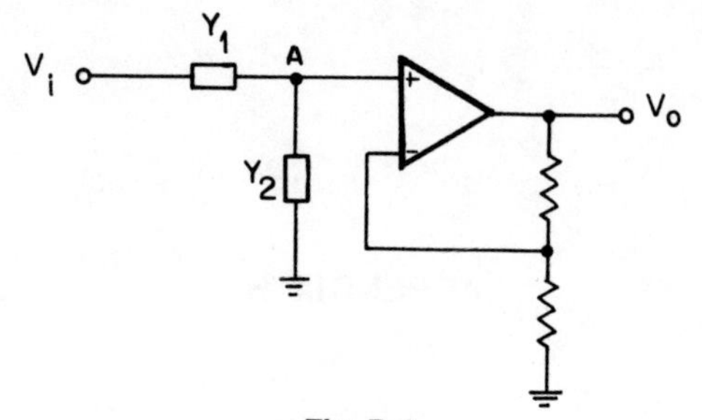

Fig. B-1.

$$Y_1 = 1/R_1$$
$$Y_2 = sC_2$$

where $s = j\omega$ and $\omega = 2\pi f$. Equation A-3 then becomes

$$\frac{V_o}{V_i} = \frac{G}{(R_1C_2)(s + 1/R_1C_2)} \quad \text{(Eq. A-5)}$$

Comparing terms of Equations A-4 and A-5 shows that

$$\omega_c = \frac{1}{R_1C_2} \quad \text{(Eq. A-6)}$$

For $\omega_2 = 1$ rad/s, Equation A-4 is normalized so that

$$\frac{V_o}{V_i} = \frac{G}{s + 1} \quad \text{(Eq. A-7)}$$

which implies that

$$R_1C_2 = 1 \quad \text{(Eq. A-8)}$$

For a High-Pass Filter

The transfer function describing a 1st order high-pass filter with a passband gain G is given by

$$\frac{V_o}{V_i} = \frac{Gs}{s + \omega_c} \quad \text{(Eq. A-9)}$$

Comparing the terms of Equations A-3 and A-9 implies that

$$Y_1 = sC_1$$
$$Y_2 = 1/R_2$$

so that Equation A-3 becomes

$$\frac{V_o}{V_i} = \frac{sGC_1}{sC_1 + 1/R_2}$$
$$= \frac{Gs}{s + 1/R_2C_1} \quad \text{(Eq. A-10)}$$

which also implies that

$$\omega_c = 1/R_2C_1 \qquad \text{(Eq. A-11)}$$

For $\omega_c = 1$ rad/s, Equation A-9 is normalized so that

$$\frac{V_o}{V_i} = \frac{Gs}{s+1} \qquad \text{(Eq. A-12)}$$

and

$$R_2C_1 = 1 \qquad \text{(Eq. A-13)}$$

B. 2ND ORDER VCVS (SALLEN-KEY) FILTER (FIG. B-2)

General Formula

At nodes A and B, the current equations are

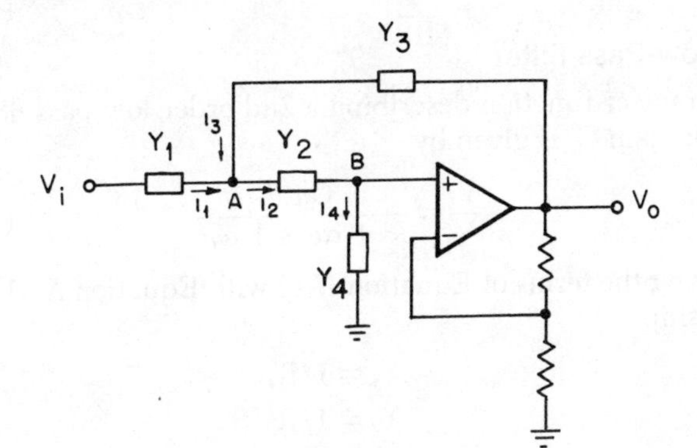

Fig. B-2.

$$I_1 + I_3 - I_2 = 0 \qquad \text{(Eq. B-1)}$$

and

$$I_2 = I_4 \qquad \text{(Eq. B-2)}$$

Also, the current through each admittance Y, is

$$I_1 = (V_i - V_A)Y_1 \qquad \text{(Eq. B-3)}$$
$$I_2 = (V_A - V_B)Y_2 \qquad \text{(Eq. B-4)}$$
$$I_3 = (V_o - V_A)Y_3 \qquad \text{(Eq. B-5)}$$
$$I_4 = V_BY_4 \qquad \text{(Eq. B-6)}$$

For a passband gain G,

$$V_o = GV_B \qquad \text{(Eq. B-7)}$$

Therefore, from Equations B-4, B-6, and B-7,

$$(V_A - V_B)Y_2 = V_B Y_4$$
$$V_A - V_o/G = V_o Y_4/GY_2$$

or

$$V_A = \frac{V_o}{G}\left(\frac{Y_4 + Y_2}{Y_2}\right) \qquad \text{(Eq. B-8)}$$

Equation B-1 then can be rewritten as

$$V_o\left[\frac{1}{G}(Y_2 + Y_4) - \frac{Y_2}{G} + \frac{Y_3}{GY_2}(Y_2 + Y_4) - Y_3 + \frac{Y_1}{GY_2}(Y_2 + Y_4)\right] = Y_1 V_i \qquad \text{(Eq. B-9)}$$

or,

$$\frac{V_o}{V_i} = \frac{GY_1Y_2}{Y_3Y_4 + Y_2Y_3(1 - G) + Y_4(Y_1 + Y_2) + Y_1Y_2} \qquad \text{(Eq. B-10)}$$

For a Low-Pass Filter

The transfer function describing a 2nd order low-pass filter with a passband gain G, is given by

$$\frac{V_o}{V_i} = \frac{G\omega_c^2}{s^2 + \alpha\omega_c s + \omega_c^2} \qquad \text{(Eq. B-11)}$$

Comparing the terms of Equation B-10 with Equation A-11 yields the relationships

$$Y_1 = 1/R_1$$
$$Y_2 = 1/R_2$$
$$Y_3 = sC_3$$
$$Y_4 = sC_4$$

Substitution of these admittances into Equation B-10, gives, after simplification,

$$\frac{V_o}{V_i} = \frac{G(1/R_1R_2C_3C_4)}{s^2 + s[1 - G/(R_2C_4) + (R_1 + R_2)/(R_1R_2C_3)] + 1/R_1R_2C_3C_4} \qquad \text{(Eq. B-12)}$$

Comparing like terms of Equations B-11 and B-12 gives

$$\omega_c^2 = \frac{1}{R_1R_2C_3C_4} \qquad \text{(Eq. B-13)}$$

and

$$\alpha\omega_c = \frac{R_1C_3(1 - G) + C_4(R_1 + R_2)}{R_1R_2C_3C_4} \qquad \text{(Eq. B-14)}$$

For Equation B-11 normalized for $\omega_c = 1$ rad/s,

$$\frac{V_o}{V_i} = \frac{G}{s^2 + \alpha s + 1} \qquad \text{(Eq. B-15)}$$

so that

$$R_1R_2C_3C_4 = 1 \qquad \text{(Eq. B-16)}$$

and

$$\alpha = \frac{R_1C_3(1 - G) + C_4(R_1 + R_2)}{R_1R_2C_3C_4} \qquad \text{(Eq. B-17)}$$

1. For $\omega_c = 1$ and $R_1 = R_2 = 1\ \Omega$, then

$$C_3 = 1/C_4 \qquad \text{(Eq. B-18)}$$

where

$$C_4 = \frac{\alpha \pm \sqrt{(\alpha^2 - 8[1 - G])}}{4} \qquad \text{(Eq. B-19)}$$

If the passband gain is unity ($G = 1$), then

$$C_4 = \alpha/2 \qquad \text{(Eq. B-20)}$$

and

$$C_3 = 2/\alpha \qquad \text{(Eq. B-21)}$$

2. For an "equal component" VCVS filter, normalized to $\omega_c = 1$ rad/s,

$$R_1 = R_2 = 1\ \Omega$$
$$C_3 = C_4 = 1\ \text{F}$$

so that Equation B-14 reduces to

$$G = 3 - \alpha \qquad \text{(Eq. B-22)}$$

Consequently, we can have only one fixed voltage gain for a given value of damping (α). For a 2nd order Butterworth response, $\alpha = 1.414$ and the gain is equal to 1.586.

For a High-Pass Filter

The transfer function describing a 2nd order high-pass filter is given by

$$\frac{V_o}{V_i} = \frac{Gs^2}{s^2 + \alpha\omega_c s + \omega_c^2} \qquad \text{(Eq. B-23)}$$

Comparing the terms of Equations B-10 and B-23 yields the relationships

$$Y_1 = sC_1$$
$$Y_2 = sC_2$$

$$Y_3 = 1/R_3$$
$$Y_4 = 1/R_4$$

Substitution of these admittances into Equation B-10 gives, after simplification

$$\frac{V_o}{V_i} = \frac{Gs^2}{s^2 + s[1 - G/(R_3C_1) + (1/C_1 + 1/C_2)/R_4] + 1/R_3R_4C_1C_2} \quad \text{(Eq. B-24)}$$

Comparing like terms of Equations B-23 and B-24,

$$\omega_c^2 = 1/R_3R_4C_1C_2 \quad \text{(Eq. B-25)}$$
$$\alpha\omega_c = 1 - G/(R_3C_1) + (1/C_1 + 1/C_2)/R_4 \quad \text{(Eq. B-26)}$$

For Equation B-23 normalized to $\omega_c = 1$ rad/s,

$$\frac{V_o}{V_i} = \frac{Gs^2}{s^2 + \alpha s + 1} \quad \text{(Eq. B-27)}$$

1. For $\omega_c = 1$ and $C_1 = C_2 = 1$ F, then

$$R_3 = 1/R_4 \quad \text{(Eq. B-28)}$$

where

$$R_4 = \frac{\alpha \pm \sqrt{(\alpha^2 - 8[1 - G])}}{2(1 - G)} \quad \text{(Eq. B-29)}$$

If the passband gain is unity, then

$$R_4 = 2/\alpha \quad \text{(Eq. B-30)}$$

and

$$R_3 = \alpha/2 \quad \text{(Eq. B-31)}$$

2. For an "equal component" VCVS filter, normalized to $\omega_c = 1$ rad/s,

$$R_3 = R_4 = 1\,\Omega$$
$$C_1 = C_2 = 1\text{ F}$$

so that Equation B-26 reduces to

$$G = 3 - \alpha \quad \text{(Eq. B-32)}$$

For a Butterworth response, $G = 1.586$.

C. 2ND ORDER MULTIPLE-FEEDBACK FILTER (FIG. B-3)

General Formula

At node A,

$$I_1 + I_4 = I_2 + I_3 \quad \text{(Eq. C-1)}$$

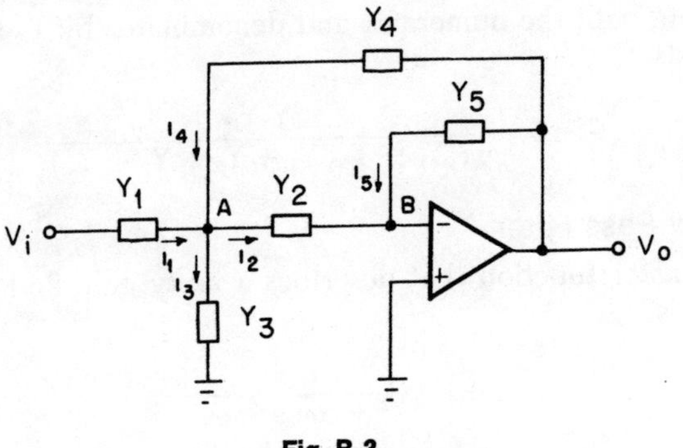

Fig. B-3.

where

$$I_1 = (V_i - V_A)Y_1 \quad \text{(Eq. C-2)}$$
$$I_2 = V_A Y_2 \quad \text{(Eq. C-3)}$$
$$I_3 = V_A Y_3 \quad \text{(Eq. C-4)}$$
$$I_4 = (V_o - V_A)Y_4 \quad \text{(Eq. C-5)}$$

Since for the ideal op-amp, no current flows into the inverting input,

$$I_2 + I_5 = 0 \quad \text{(Eq. C-6)}$$

and since

$$V_o Y_5 = I_5 \quad \text{(Eq. C-7)}$$

then Equation C-3 reduces to

$$V_A = -\frac{V_o Y_5}{Y_2} \quad \text{(Eq. C-8)}$$

Substituting Equations C-2 through C-5 into Equation C-1, we obtain

$$V_i Y_1 - V_A Y_1 + V_o Y_4 - V_A Y_4 = V_A Y_2 + V_A Y_3 \quad \text{(Eq. C-9)}$$

and substitution of Equation C-8 into Equation C-9 gives

$$V_i Y_1 + V_o(Y_1 Y_5/Y_2) + V_o Y_4 + V_o(Y_4 Y_5/Y_2) = -V_o Y_5 - V_o(Y_3 Y_5/Y_2)$$

Rearranging terms,

$$-V_o\left(\frac{Y_1 Y_5}{Y_2} + Y_4 + \frac{Y_4 Y_5}{Y_2} + Y_5 + \frac{Y_3 Y_5}{Y_2}\right) = V_i Y_1$$

or,

$$\frac{V_o}{V_i} = -\frac{Y_1}{\dfrac{Y_1 Y_5}{Y_2} + Y_4 + \dfrac{Y_4 Y_5}{Y_2} + Y_5 + \dfrac{Y_3 Y_5}{Y_2}}$$

Multiplying both the numerator and denominator by Y_2 and simplifying yields

$$\frac{V_o}{V_i} - = -\frac{Y_1 Y_2}{Y_2 Y_4 + Y_5(Y_1 + Y_2 + Y_3 + Y_4)} \qquad \text{(Eq. C-10)}$$

For a Low-Pass Filter

The transfer function that describes a unity gain 2nd order low-pass filter is

$$\frac{V_o}{V_i} = \frac{\omega_c^2}{s^2 + \alpha\omega_c s + \omega_c^2} \qquad \text{(Eq. C-11)}$$

Comparing the terms of Equation C-10 with Equation C-11 yields the relationships

$$Y_1 = 1/R_1$$
$$Y_2 = 1/R_2$$
$$Y_3 = sC_3$$
$$Y_4 = 1/R_4$$
$$Y_5 = sC_5$$

Substitution of these admittances into Equation C-10 after simplifying gives

$$\frac{V_o}{V_i} = \frac{-(1/R_1R_2C_3C_5)}{s^2 + s[(1/R_1 + 1/R_2 + 1/R_4)/C_3] + 1/R_2R_4C_3C_5} \qquad \text{(Eq. C-12)}$$

Comparing like terms of Equations C-11 and C-12 gives

$$\omega_c^2 = \frac{1}{R_1R_2C_3C_5}$$

(Eq. C-13)

or

$$= \frac{1}{R_2R_4C_3C_5}$$

and

$$\alpha\omega_c = (1/R_1 + 1/R_2 + 1/R_4)/C_3 \qquad \text{(Eq. C-14)}$$

Normalizing Equation C-11 for $\omega_c = 1$ rad/s,

$$\frac{V_o}{V_i} = \frac{-1}{s^2 + \alpha s + 1} \qquad \text{(Eq. C-15)}$$

so that

$$R_1R_2C_3C_5 = R_2R_4C_3C_5 \qquad \text{(Eq. C-16)}$$

and

$$\alpha = (1/R_1 + 1/R_2 + 1/R_4)/C_3 \qquad \text{(Eq. C-17)}$$

1. For $\omega_c = 1$ and $R_1 = R_2 = 1\Omega$, then

$$R_4 = R_1 \qquad \text{(Eq. C-18)}$$

$$C_3 = 1/C_5 \qquad \text{(Eq. C-19)}$$

where

$$C_5 = \alpha/3 \qquad \text{(Eq. C-20)}$$

2. For $\omega_c = 1$ and $R_1 = 2R_2 = 1\Omega$,

$$R_4 = R_1 \qquad \text{(Eq. C-21)}$$

$$R_2 = R_1/2 \qquad \text{(Eq. C-22)}$$

$$C_5 = 2/C_3 \qquad \text{(Eq. C-23)}$$

where

$$C_3 = 4/\alpha \qquad \text{(Eq. C-24)}$$

For a High-Pass Filter

The transfer function that describes a unity gain 2nd order high-pass filter is

$$\frac{V_o}{V_i} = \frac{-s^2}{s^2 + \alpha\omega_c s + \omega_c^2} \qquad \text{(Eq. C-25)}$$

Comparing the terms of Equation C-10 with Equation C-25 yields the relationships

$$Y_1 = sC_1$$

$$Y_2 = sC_2$$

$$Y_3 = 1/R_3$$

$$Y_4 = sC_4$$

$$Y_5 = 1/R_5$$

Substitution of these admittances into Equation C-10 after simplifying gives

$$\frac{V_o}{V_i} = \frac{-s^2}{s^2(C_4/C_1) + s[(C_1 + C_2 + C_4)/R_5] + 1/R_3R_5C_1C_2} \qquad \text{(Eq. C-26)}$$

Comparing like terms of Equations C-25 and C-26 gives

$$C_4/C_1 = 1 \qquad \text{(Eq. C-27)}$$

$$\omega_c^2 = 1/R_3R_5C_1C_2 \qquad \text{(Eq. C-28)}$$

and

$$\alpha\omega_c = (C_1 + C_2 + C_4)/R_5 \qquad \text{(Eq. C-29)}$$

For Equation C-25 normalized for $\omega_c = 1$ rad/s,

$$\frac{V_o}{V_i} = \frac{-s^2}{s^2 + \alpha s + 1} \qquad \text{(Eq. C-30)}$$

1. For $\omega_c = 1$, and $C_1 = C_2 = C_4 = 1$ F, then

$$R_3 = 1/R_5 \qquad \text{(Eq. C-31)}$$

where

$$R_5 = 3/\alpha \qquad \text{(Eq. C-32)}$$

2. For $\omega_c = 1$, and $C_1 = 2C_2 = 1$ F, then

$$C_4 = C_1 \qquad \text{(Eq. C-33)}$$

$$C_2 = C_1/2 \qquad \text{(Eq. C-34)}$$

$$R_3 = 2/R_5 \qquad \text{(Eq. C-35)}$$

where

$$R_5 = 5/2\alpha \qquad \text{(Eq. C-36)}$$

For a Bandpass Filter (1 pole)

The transfer function that describes a 2nd order (1 pole) bandpass filter is

$$\frac{V_o}{V_i} = \frac{-\omega_o s}{s^2 + (\omega_o/Q)s + \omega_o^2} \qquad \text{(Eq. C-37)}$$

where $\alpha = 1/Q$. Comparing terms of equations C-10 with C-37 yields the relationships

$$Y_1 = 1/R_1$$

$$Y_2 = sC_2$$

$$Y_3 = 1/R_3$$

$$Y_4 = sC_4$$

$$Y_5 = 1/R_5$$

so that Equation C-10 is now written as

$$\frac{V_o}{V_i} = \frac{-s(1/R_1C_4)}{s^2 + s[(C_2 + C_4)/R_5C_2C_4] + (1/R_1 + 1/R_3)/R_5C_2C_4} \qquad \text{(Eq. C-38)}$$

Comparing like terms of Equation C-37 with Equation C-38 gives

$$\omega_o = 1/R_1C_4 \quad \text{(Eq. C-39)}$$

$$\omega_o^2 = (1/R_1 + 1/R_3)/R_5C_2C_4 \quad \text{(Eq. C-40)}$$

and

$$\omega_o/Q = (C_2 + C_4)/R_5C_2C_4 \quad \text{(Eq. C-41)}$$

For Equation C-37 normalized for $\omega_o = 1$ rad/s,

$$\frac{V_o}{V_i} = \frac{-s}{s^2 + s/Q + 1} \quad \text{(Eq. C-42)}$$

1. For $\omega_o = 1$, and $C_2 = C_4 = 1$ F, then

$$R_1 = 1 \quad \text{(Eq. C-43)}$$

$$R_5 = 2Q \quad \text{(Eq. C-44)}$$

$$R_3 = 1/(2Q - 1) \quad \text{(Eq. C-45)}$$

Substituting Equations C-43, C-44, and C-45 into Equation C-38 for $\omega_o = 1$, so that $s = 1$ and $s^2 = -1$, we then obtain the voltage gain such that

$$\frac{V_o}{V_i} = -Q \quad \text{(Eq. C-46)}$$

2. For $\omega_o = 1$, $R_3 = \infty$ $(Y_3 = 0)$, and $C_2 = C_4 = 1$ F, then

$$R_1 = 1/R_5 \quad \text{(Eq. C-47)}$$

where

$$R_5 = 2Q \quad \text{(Eq. C-48)}$$

Substituting Equations C-47 and C-48 into Equation C-38 for $\omega_o = 1$, we obtain the voltage gain such that

$$\frac{V_o}{V_i} = -2Q^2 \quad \text{(Eq. C-49)}$$

D. STATE-VARIABLE FILTERS (FIG. B-4)

General Expressions:

From the circuit in Fig. B-4, the following three equations can be written:

$$V_{HP} = -GV_i - V_{LP} + MV_{BP} \quad \text{(Eq. D-1)}$$

$$V_{BP} = \int V_{HP}\,dt = -\frac{V_{HP}}{sR_1C_1} \quad \text{(Eq. D-2)}$$

$$V_{LP} = \int V_{BP}\,dt = -\frac{V_{BP}}{sR_2C_2} \quad \text{(Eq. D-3)}$$

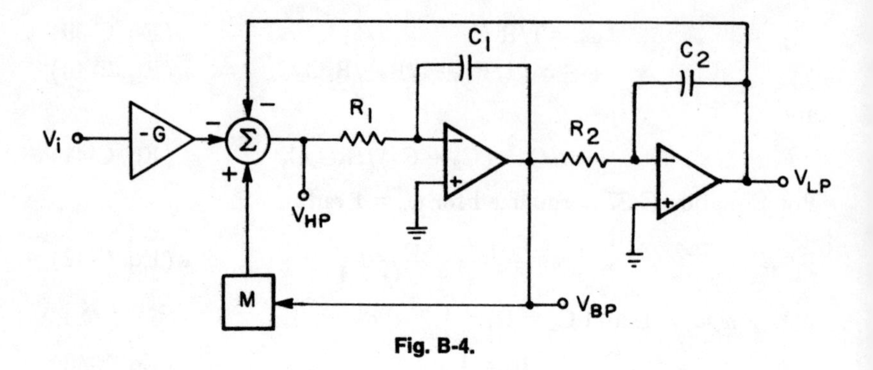

Fig. B-4.

For a Low-Pass Filter

The output voltage will be, according to Equation D-3,

$$V_o = V_{LP}$$

$$= -\frac{V_{BP}}{sR_2C_2} \qquad \text{(Eq. D-4)}$$

After solving for V_{BP}, and using Equations D-1 and D-2, we obtain

$$V_o[s^2(R_1R_2C_1C_2) + Ms(R_2C_2) + 1] = -GV_i$$

or

$$\frac{V_o}{V_i} = -\frac{G}{s^2(R_1R_2C_1C_2) + Ms(R_2C_2) + 1}$$

$$= -\frac{G/R_1R_2C_1C_2}{s_2 + (M/R_1C_1)s + 1/R_1R_2C_1C_2} \qquad \text{(Eq. D-5)}$$

Comparing the terms of the above equation with Equation B-11 yields the following relationships:

$$\omega_c{}^2 = \frac{1}{R_1R_2C_1C_2} \qquad \text{(Eq. D-6)}$$

$$\omega_c = \frac{1}{R_1C_1} \qquad \text{(Eq. D-7)}$$

which implies that $R_1C_1 = R_2C_2$, and

$$\alpha = M \qquad \text{(Eq. D-8)}$$

For a High-Pass Filter

From Equation D-2, the output voltage is

$$V_o = V_{HP}$$

$$= -GV_i - V_{LP} + MV_{BP} \qquad \text{(Eq. D-9)}$$

After solving for V_{LP}, and using Equations D-1 and D-3, we obtain

$$V_o[1 + 1/(s^2R_1R_2C_1C_2) + M/sR_1C_1] = -GV_i$$

or

$$\frac{V_o}{V_i} = -\frac{Gs^2(R_1R_2C_1C_2)}{s^2(R_1R_2C_1C_2) + Ms(R_2C_2) + 1}$$

$$= -\frac{Gs^2}{s^2 + (M/R_1C_1)s + 1/R_1R_2C_1C_2} \quad \text{(Eq. D-10)}$$

Comparing the terms of the above equation with Equation B-23, we obtain the following relationships:

$$\omega_c^2 = \frac{1}{R_1R_2C_1C_2} \quad \text{(Eq. D-11)}$$

$$\omega_c = \frac{1}{R_1C_1} \quad \text{(Eq. D-12)}$$

which implies that $R_1C_1 = R_2C_2$, and

$$\alpha = M \quad \text{(Eq. D-13)}$$

For a Bandpass Filter

Solving Equations D-1, D-2, and D-3 simultaneously for $V_o = V_{BP}$ in terms of V_i, we obtain

$$V_o[(1/sR_2C_2) + s(R_1C_1) + M] = GV_i$$

or

$$\frac{V_o}{V_i} = \frac{Gs(R_2C_2)}{s^2(R_1R_2C_1C_2) + Ms(R_2C_2) + 1}$$

$$= \frac{Gs/R_1C_1}{s^2 + Ms(1/R_1C_1) + 1/R_1R_2C_1C_2} \quad \text{(Eq. D-14)}$$

Comparing the terms of the above equation with Equation C-37 (multiplied by a factor of G), we obtain the relationships:

$$\omega_o^2 = 1/R_1R_2C_1C_2 \quad \text{(Eq. D-15)}$$

$$R_1C_1 = R_2C_2 \quad \text{(Eq. D-16)}$$

and

$$M = \frac{1}{Q} \text{ (or } \alpha) \quad \text{(Eq. D-17)}$$

Using a Fixed-Gain Summing Block (Fig. B-5)

The input-output relationships can be expressed as

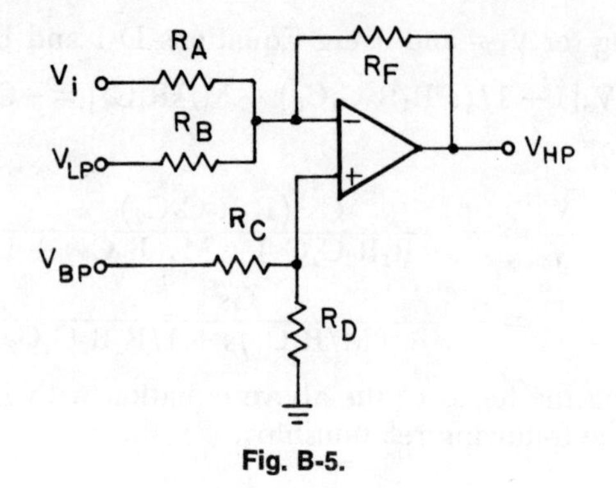

Fig. B-5.

$$\frac{V_{HP}}{V_i} = -\frac{R_F}{R_A} = G_A \qquad \text{(Eq. D-18)}$$

$$\frac{V_{HP}}{V_{LP}} = -\frac{R_F}{R_B} = G_B \qquad \text{(Eq. D-19)}$$

and

$$\frac{V_{HP}}{V_{BP}} = \left[\frac{R_F R_A + R_F R_B + R_A R_B}{R_A + R_B}\right]\left[\frac{R_D}{R_C + R_D}\right] = G_C \qquad \text{(Eq. D-20)}$$

By summing the three inputs,

$$V_{HP} = -G_A V_i - G_B V_{LP} + G_C V_{BP} \qquad \text{(Eq. D-21)}$$

which is the same form as Equation D-1, or

$$V_{HP} = -GV_i - V_{LP} + MV_{BP}$$

implying the equalities:

$$G_A = G = 1$$
$$G_B = 1$$
$$G_C = M = \alpha$$

Consequently, $R_A = R_B = R_F = 1\Omega$ (for convenience). To determine R_C, first let $R_D = 1\Omega$, so that equation D-20 becomes

$$G_C = 3\left(\frac{1}{R_C + 1}\right) = \alpha \qquad \text{(Eq. D-22)}$$

Solving for R_C in terms of α, we obtain

$$R_C = \frac{3 - \alpha}{\alpha} \qquad \text{(Eq. D-23)}$$

If we are interested in a Butterworth response, $\alpha = 1.414$, and R_C then equals 1.121Ω.

APPENDIX C

Butterworth Damping Coefficients

The polynomial

$$P(s) = a_n s^n + a_{n-1} s^{n-1} + \ldots + a_1 s + a_o \qquad \text{(Eq. E-1)}$$

where the coefficients are real and all roots have negative real parts, so that $P(s)$ is related to the Butterworth function

$$H(s) = \frac{1}{P(s)} \qquad \text{(Eq. E-2)}$$

which can be factored into the form

$$H(s) = \frac{1}{(s + s_n)(s + s_{n-1})(s + s_{n-2}) \ldots (s + s_1)} \qquad \text{(Eq. E-3)}$$

For a Butterworth response, the poles of Equation E-3 are given by

$$s_k = e^{j[(2k+n-1)/2n]\pi}, \quad k = 1, 2, 3 \ldots, 2n \qquad \text{(Eq. E-4)}$$

Expressing s_k as a complex number,

$$s_k = \sigma_k \pm j\omega_k \qquad \text{(Eq. E-5)}$$

then

$$\sigma_k = \sin \left[\frac{2k-1}{2n}\right] \pi \qquad \text{(Eq. E-6)}$$

$$\omega_k = \cos \left[\frac{2k-1}{2n}\right] \pi \qquad \text{(Eq. E-7)}$$

1. For a 1st order filter, n = 1, and

$$k = 1: \quad \sigma_1 = \sin[\pi/2] = 1 \quad \omega_1 = \cos[\pi/2] = 0$$

so that equation E-3 becomes

$$H_1(s) = \frac{1}{(s+1)} \qquad \text{(Eq. E-8)}$$

2. For a 2nd order filter,

$$k = 1: \quad \sigma_1 = \sin[\pi/4] = 0.707 \quad \omega_1 = \cos[\pi/4] = 0.707$$
$$k = 2: \quad \sigma_2 = \sin[3\pi/4] = 0.707 \quad \omega_2 = \cos[3\pi/4] = -0.707$$

so that Equation E-3 becomes

$$H_2(s) = \frac{1}{(s + 0.707 + j0.707)(s + 0.707 - j0.707)}$$

or

$$= \frac{1}{s^2 + 1.414s + 1} \qquad \text{(Eq. E-9)}$$

3. For a 3rd order filter,

$$k = 1: \quad \sigma_1 = \sin[\pi/6] = 0.500 \quad \omega_1 = \cos[\pi/6] = 0.866$$
$$k = 2: \quad \sigma_2 = \sin[\pi/2] = 1.000 \quad \omega_2 = \cos[\pi/2] = 0$$
$$k = 3: \quad \sigma_3 = \sin[5\pi/6] = 0.500 \quad \omega_3 = \cos[5\pi/6] = -0.866$$

so that Equation E-3 becomes

$$H_3(s) = \frac{1}{(s+1)(s + 0.5 + j0.866)(s + 0.5 - j0.866)}$$

or

$$= \frac{1}{(s+1)(s^2 + s + 1)} \qquad \text{(Eq. E-10)}$$

4. For a 4th order filter,

$$k = 1: \quad \sigma_1 = \sin[\pi/8] = 0.3827 \quad \omega_1 = \cos[\pi/8] = 0.9239$$
$$k = 2: \quad \sigma_2 = \sin[3\pi/8] = 0.9239 \quad \omega_2 = \cos[3\pi/8] = 0.3827$$
$$k = 3: \quad \sigma_3 = \sin[5\pi/8] = 0.9239 \quad \omega_3 = \cos[5\pi/8] = -0.3827$$
$$k = 4: \quad \sigma_4 = \sin[7\pi/8] = 0.3827 \quad \omega_4 = \cos[7\pi/8] = -0.9239$$

so that Equation E-3 is

$$H_4(s) = \frac{1}{(s + s_1)(s + s_2)(s + s_3)(s + s_4)}$$

where, for example,

$$s_1 = 0.3827 + j0.9239$$

Then $H_4(s)$ can be written as

$$H_4(s) = \frac{1}{(s^2 + 1.8478s + 1)(s^2 + 0.7654s + 1)} \qquad \text{(Eq. E-11)}$$

5. Applying the same method for higher order filters,

$$H_5(s) = \frac{1}{(s + 1)(s^2 + 0.6180s + 1)(s^2 + 1.6180s + 1)} \qquad \text{(Eq. E-12)}$$

$$H_6(s) = \frac{1}{(s^2 + 0.5176s + 1)(s^2 + 1.412s + 1)(s^2 + 1.9318s + 1)} \qquad \text{(Eq. E-13)}$$

The required damping coefficients (the coefficients of all s terms in the denominator of H[s]) are summarized in the following table up to a 6th order filter.

Order	*1st Section*	*2nd Section*	*3rd Section*
1	1.0000		
2	1.4142		
3	1.0000	1.0000	
4	1.8478	0.7654	
5	1.6180	0.6180	1.000
6	1.9318	1.4142	0.5176

For Equations E-8 through E-13, by letting

$$s = j\omega$$
$$s^2 = -\omega^2$$
$$s^3 = -j\omega^3$$
$$s^4 = \omega^4$$
$$s^5 = j\omega^5$$
$$s^6 = -\omega^6$$

it can be shown that the amplitude response of any nth order Butterworth filter can be expressed in the form of

$$dB = 20 \log_{10} \frac{1}{[1 + (\omega)^{2n}]^{1/2}} \qquad \text{(Eq. E-14)}$$

where n is the order of the filter. For example, using Equation E-10 (a 3rd order filter),

$$\frac{1}{(s+1)(s^2+s+1)} = \frac{1}{(s^3+2s^2+2s+1)}$$

$$= \frac{1}{(1-2\omega^2)+j(2\omega-\omega^3)}$$

Then the magnitude is

$$20 \log_{10} \left| \frac{1}{(s+1)(S^2+s+1)} \right| = 20 \log_{10}$$

$$\left| \frac{1}{[(1-2\omega^2)^2+(2\omega-\omega^3)^2]^{1/2}} \right|$$

$$= 20 \log_{10} \frac{1}{(1-4\omega^2+4\omega^4+4\omega^2-4\omega^4+\omega^6)^{1/2}}$$

$$= 20 \log_{10} \frac{1}{(1+\omega^6)^{1/2}}$$

which is exactly Equation E-14 for $n = 3$.

APPENDIX D

Parts List for Experiments

CHAPTER 1

Experiment No. 1

Quantity	*Description*
1	SK-10 breadboarding socket
1	Dual polarity power supply
1	LR-31 Signal Generator Outboard *or* function generator
1	Oscilloscope
1	0.1 μF capacitor (used with LR-31)
–	Hookup wire

Experiment No. 2

Quantity	*Description*
1	SK-10 breadboarding socket
1	Dual polarity power supply
1	LR-31 Signal Generator Outboard *or* function generator
1	Oscilloscope
1	0.1 μF capacitor (used with LR-31)
–	Hookup wire

CHAPTER 2

Experiment No. 1

Quantity	*Description*
1	SK-10 breadboarding socket
1	Dual polarity power supply
1	LR-31 Signal Generator Outboard *or* function generator
1	741 op amp

1	0.1 μF capacitor (used with LR-31)
2	10 kΩ resistor
1	15 kΩ resistor
1	27 kΩ resistor
1	47 kΩ resistor
1	100 kΩ resistor
1	Dual trace oscilloscope
–	Hookup wire

Experiment No. 2

Quantity	*Description*
1	SK-10 breadboarding socket
1	Dual polarity power supply
1	LR-31 Signal Generator Outboard *or* function generator
1	741 op amp
1	0.1 μF capacitor (used with LR-31)
2	10 kΩ resistor
1	27 kΩ resistor
1	39 kΩ resistor
1	47 kΩ resistor
1	100 kΩ resistor
1	Dual trace oscilloscope
–	Hookup wire

Experiment No. 3

Quantity	*Description*
1	SK-10 breadboarding socket
1	Dual polarity power supply
1	LR-31 Signal Generator Outboard *or* function generator
1	741 op amp
1	0.1 μF capacitor (used with LR-31)
1	Dual trace oscilloscope
–	Hookup wire

Experiment No. 4

Quantity	*Description*
1	SK-10 breadboarding socket
1	Dual polarity power supply
1	LR-31 Signal Generator Outboard *or* function generator
2	741 op amp
1	0.1 μF capacitor (use with LR-31)
3	10 kΩ resistor
2	27 kΩ resistor
1	Dual trace oscilloscope
–	Hookup wire

Experiment No. 5

Quantity	*Description*
1	SK-10 breadboarding socket
1	Dual polarity power supply
1	LR-31 Signal Generator Outboard *or* function generator
1	741 op amp
1	0.1 μF capacitor (used with LR-31)
1	0.01 μF capacitor
1	1 kΩ resistor
2	10 kΩ resistor
2	100 kΩ resistor
1	Dual trace oscilloscope
–	Hookup wire

CHAPTER 4

Experiment No. 1

Quantity	*Description*
1	SK-10 breadboarding socket
1	Dual polarity power supply
1	LR-31 Signal Generator Outboard *or* function generator
1	741 op amp
1	0.1 μF capacitor (used with LR-31)
1	0.01 μF capacitor
1	0.033 μF capacitor
1	4.7 kΩ resistor
1	Dual trace oscilloscope
–	Hookup wire

Experiment No. 2

Quantity	*Description*
1	SK-10 breadboarding socket
1	Dual polarity power supply
1	LR-31 Signal Generator Outboard *or* function generator
1	741 op amp
1	0.1 μF capacitor (used with LR-31)
1	0.01 μF capacitor
1	0.033 μF capacitor
1	4.7 kΩ resistor
2	10 kΩ resistor
1	Dual trace oscilloscope

Experiment No. 3

Quantity	*Description*
1	SK-10 breadboarding socket
1	Dual polarity power supply
1	LR-31 Signal Generator Outboard *or* function generator
1	741 op amp
1	0.1 μF capacitor (used with LR-31)
2	0.033 μF capacitor
2	4.7 kΩ resistor
2	10 kΩ resistor
1	Dual trace oscilloscope
–	Hookup wire

Experiment No. 4

Parts list identical to Experiment No. 1 of this chapter

CHAPTER 5

Experiment No. 1

Quantity	*Description*
1	SK-10 breadboarding socket
1	Dual polarity power supply
1	LR-31 Signal Generator Outboard *or* function generator
1	741 op amp
1	0.1 μF capacitor (used with LR-31)
3	0.033 μF capacitor
2	3.3 kΩ resistor
1	Dual trace oscilloscope
–	Hookup wire

Experiment No. 2

Quantity	*Description*
1	SK-0 breadboarding socket
1	Dual polarity power supply
1	LR-31 Signal Generator Outboard *or* function generator
1	741 op amp
1	0.1 μF capacitor (used with LR-31)
2	0.033 μF capacitor
2	4.7 kΩ resistor
1	27 kΩ resistor
1	47 kΩ resistor
1	Dual trace oscilloscope
–	Hookup wire

CHAPTER 6

Experiment No. 1

Quantity	*Description*
1	SK-10 breadboarding socket
1	Dual polarity power supply
1	LR-31 Signal Generator Outboard *or* function generator
1	741 op amp
2	0.1 μF capacitor (1 used with LR-31)
1	0.022 μF capacitor
3	6.8 kΩ resistor
1	Dual trace oscilloscope
–	Hookup wire

Experiment No. 2

Quantity	*Description*
1	SK-10 breadboarding socket
1	Dual polarity power supply
1	LR-31 Signal Generator Outboard *or* function generator
1	741 op amp
1	0.1 μF capacitor (used with LR-31)
3	0.033 μF capacitor
1	2.2 kΩ resistor
1	10 kΩ resistor
1	Dual trace oscilloscope
–	Hookup wire

CHAPTER 7

Experiment No. 1

Quantity	*Description*
1	SK-10 breadboarding socket
1	Dual polarity power supply
1	LR-31 Signal Generator Outboard *or* function generator
2	741 op amp
1	0.1 μF capacitor (used with LR-31)
3	0.033 μF capacitor
3	4.7 kΩ resistor
2	10 kΩ resistor
1	Dual trace oscilloscope
–	Hookup wire

Experiment No. 2

Quantity	*Description*
1	SK-10 breadboarding socket
1	Dual polarity power supply

1	LR-31 Signal Generator Outboard *or* function generator
2	741 op amp
1	0.1 μF capacitor (used with LR-31)
4	0.01 μF capacitor
1	2.2 kΩ resistor
4	8.2 kΩ resistor
1	15 kΩ resistor
1	18 kΩ resistor
1	22 kΩ resistor
1	Dual trace oscilloscope
–	Hookup wire

Experiment No. 3

Quantity	*Description*
1	SK-10 breadboarding socket
1	Dual polarity power supply
1	LR-31 Signal Generator Outboard *or* function generator
3	741 op amp
1	0.1 μF capacitor (used with LR-31)
5	0.033 μF capacitor
5	10 kΩ resistor
1	12 kΩ resistor
1	15 kΩ resistor
1	18 kΩ resistor
1	39 kΩ resistor
1	Dual trace oscilloscope
–	Hookup wire

CHAPTER 8

Experiment No. 1

Quantity	*Description*
1	SK-10 breadboarding socket
1	Dual polarity power supply
1	LR-31 Signal Generator Outboard *or* function generator
1	741 op amp
1	0.1 μF capacitor (used with LR-31)
2	0.01 μF capacitor
1	1.5 kΩ resistor
1	2.7 kΩ resistor
1	68 kΩ resistor
1	180 kΩ resistor
1	Dual trace oscilloscope
–	Hookup wire

Experiment No. 2

Quantity	*Description*
1	SK-10 breadboarding socket
1	Dual polarity power supply
1	LR-31 Signal Generator Outboard *or* function generator
2	741 op amp
1	0.1 μF capacitor (used with LR-31)
4	0.01 μF capacitor
2	2.7 kΩ resistor
2	68 kΩ resistor
2	180 kΩ resistor
1	Dual trace oscilloscope
–	Hookup wire

Experiment No. 3

Quantity	*Description*
1	SK-10 breadboarding socket
1	Dual polarity power supply
1	LR-31 Signal Generator Outboard *or* function generator
2	741 op amp
1	0.1 μF capacitor (used with LR-31)
2	0.047 μF capacitor
2	0.01 μF capacitor
2	5.6 kΩ resistor
2	8.2 kΩ resistor
2	27 kΩ resistor
2	47 kΩ resistor
1	Dual trace oscilloscope
–	Hookup wire

Experiment No. 4

Quantity	*Description*
1	SK-10 breadboarding socket
1	Dual polarity power supply
1	LR-31 Signal Generator Outboard *or* function generator
1	741 op amp
1	0.1 μF capacitor (used with LR-31)
4	0.0047 μF capacitor
1	0.001 μF capacitor
4	33 kΩ resistor
2	220 kΩ resistor
1	Dual trace oscilloscope
–	Hookup wire

Experiment No. 5

Quantity	*Description*
1	SK-10 breadboarding socket
1	Dual polarity power supply
1	LR-31 Signal Generator Outboard *or* function generator
2	741 op amp
1	0.1 μF capacitor (used with LR-31)
2	0.01 μF capacitor
1	2.7 kΩ resistor
2	6.8 kΩ resistor
1	10 kΩ resistor
1	68 kΩ resistor
1	180 kΩ resistor
1	Dual trace oscilloscope
–	Hookup wire

CHAPTER 9

Experiment No. 1

Quantity	*Description*
1	SK-10 breadboarding socket
1	Dual polarity power supply
1	LR-31 Signal Generator Outboard *or* function generator
3	741 op amp
1	0.1 μF capacitor (used with LR-31)
2	0.022 μF capacitor
1	2.7 kΩ resistor
1	3.0 kΩ resistor
7	6.8 kΩ resistor
1	27 kΩ resistor
1	270 kΩ resistor
1	Dual trace oscilloscope
–	Hookup wire

Experiment No. 2

Quantity	*Description*
1	SK-10 breadboarding socket
1	Dual polarity power supply
1	LR-31 Signal Generator Outboard *or* function generator
4	741 op amp
1	0.1 μF capacitor (used with LR-31)
2	0.0047 μF capacitor
1	1 kΩ resistor

5	10 kΩ resistor
1	47 kΩ resistor
2	56 kΩ resistor
1	68 kΩ resistor
1	680 kΩ resistor
1	Dual trace oscilloscope
–	Hookup wire

Experiment No. 3

Quantity	*Description*
1	SK-10 breadboarding socket
1	Dual polarity power supply
1	LR-31 Signal Generator Outboard *or* function generator
5	741 op amp
1	0.1 μF capacitor (used with LR-31)
2	0.0047 μF capacitor
7	10 kΩ resistor
2	56 kΩ resistor
1	68 kΩ resistor
1	680 kΩ resistor
1	Dual trace oscilloscope
–	Hookup wire

Index

TO THE READER

This book is one of an expanding series of books that will cover the field of basic electronics and digital electronics from basic gates and flip-flops through microcomputers and digital telecommunications. We are attempting to develop a mailing list of individuals who would like to receive information on the series. We would be delighted to add your name to it if you would fill in the information below and mail this sheet to us. Thanks.

1. I have the following books:

____

____

____

____

____

____

____

____

____

____

____

____

2. My occupation is: ☐ student ☐ teacher, instructor ☐ hobbyist

☐ housewife ☐ scientist, engineer, doctor, etc. ☐ businessman

☐ Other: ______________________

Name (print): ______________________

Address ______________________

City ______________________ State ______________

Zip Code ______________________

Mail to:

Books
P.O. Box 715
Blacksburg, Virginia 24060